GROWING
PLANTS FOR FREE

Spanish iris (*I. xiphium* form)

D0870844

Geoff Bryant

CASSELL

Cassell Publishers Limited
Wellington House, 125 Strand
London WC2R 0BB

First published in Great Britain 1995
in association with
David Bateman Limited
Tarndale Grove, Albany Business Park
Bush Road, Albany
Auckland, New Zealand

Distributed in the United States by Sterling Publishing Co. Inc,
387 Park Avenue South, New York, NY 10016, USA

British Library Cataloguing in Publication Data
A catalogue record for this book is available from the British
Library

ISBN 0-304-34673-X

Printed in Hong Kong by Colorcraft Ltd

Bluebells (*Hyacinthoides*) in a
woodland setting.

Contents

INTRODUCTION

Nymphaea 'Chromatella'

AS children, most of us experienced sowing a few seeds and marvelling at the plants that resulted. Yet despite that sense of wonder, and even though it seemed so easy, many gardeners are content to buy all their plants and never try to propagate any. Why? Well, over the years plant propagation has become steeped in mystery, to the extent that anything beyond sowing a few vegetable or flower seeds is often viewed as far too complicated for the average gardener. It seems that many gardeners think plant propagation is something akin to alchemy; it's not. Propagation is easy once you have mastered the basic techniques and there is nothing to stop anyone producing plants that are at least as good as those available from any nursery.

That said, there is a world of difference between basic plant propagation and the scientific methods employed by the horticulture and forestry industries. This book is intended for home gardeners, not scientists, so although the tables around which the book is based are very comprehensive, you won't find masses of scientific data, nor will you need it. Also, there are usually several ways to propagate a plant and which is best depends on a variety of factors, such as the equipment at your disposal, your level of experience and the climate. The following pages outline the methods to follow and point out some of the pitfalls to avoid. It is largely a matter of following the rules, getting the timing right and knowing what you are doing.

Successful propagation depends on both experience and a professional attitude. Experience is most often gained on the job, but plenty of reading helps. Beginning propagators need to work at developing a comprehensive understanding of plant types, families and relationships, because the more you know about the plants that you are working with, the better your results will be. Also, if something does go wrong, you'll be more likely to understand why.

The professional attitude is a combination of pride in your work and an understanding of plant quality. Many home gardeners simply throw a few seeds in a tray of old garden soil and are quite happy to put up with whatever plants result, if any. Yet when buying plants from a nursery or garden centre, the same gardeners demand the best. You should apply the standards that you expect of a professional to your own work and have no qualms about consigning your lesser efforts to the compost heap — put them down to experience.

You may have been put off propagating plants because you thought that specialised equipment was required. While it is true that the better your equipment the better, faster and more consistent your results will be, there is no need to have elaborate gear to start with. Hundreds of different plants can be propagated with little more than a few propagating trays, a good pair of secateurs, decent potting mix and some sturdy plastic bags.

If you have a passion for plants, propagation provides the ultimate in satisfaction. The gardeners among us find that propagation provides new insights into plants as well as making gardening cheaper, but propagation is not just for gardeners. There are many plant enthusiasts, and I include myself among them, for whom gardening is little more than hard work. If you enjoy looking at and studying plants, but are unable or unwilling to spend a lot of time gardening, the chances are that you'll get a lot of satisfaction from plant propagation. It enables you to see how the relationships between plants work and gives you the chance to put all that reading into practice. If you don't have a garden you may have to give away the plants, but that can be very satisfying in itself. Whatever your reasons for wanting to propagate plants, be careful, it is addictive.

Chapter 1

WHAT IS PLANT PROPAGATION AND WHY DO IT?

Livingstone daisies

PLANT propagation is plant production, making two or more plants from one. There are two main divisions of propagation, sexual and asexual (or vegetative). Lower plants, such as mosses and ferns, reproduce naturally by means of spores, which are fertilised in the presence of water; higher plants produce flowers and set seed through pollination. These are sexual methods of reproduction. Vegetative propagation uses existing plant material to produce new plants and does not require any seed or spores to be fertilised. Layering, cuttings, grafting, budding and tissue culture are common methods of vegetative propagation.

When a plant reproduces sexually a large element of chance is introduced. A seedling is the sum of its parents; and, as with a child, no seedling is exactly like its parents. We can breed seedling strains that are very consistent and noticeably superior to their parents, but we cannot make exact replicas by seed. Even seed from naturally occurring species will produce seedlings with some very slight variations.

If there was no way to reproduce plants other than by sexual means our gardens would look very different and plant hybridising would be an infinitely more frustrating process. Vegetative propagation is the answer to these problems because it takes part of just one plant and uses it to produce many more. Every cutting is genetically an exact replica of the plant it was taken from. It contains precisely the same genes and chromosomes and is therefore a perfect copy, a clone.

Although vegetative propagation is usually carried out under controlled conditions, it would be wrong to think of it as any less natural than sexual propagation. Plants regularly reproduce asexually without any outside help. Low-growing, spreading plants, such as most ground covers, have stems that are constantly in contact with the soil. These stems often strike roots as they spread. Layering, aerial layering and cuttings adapt this natural process. The tightly wound stems of a climber often fuse after a prolonged period of contact. It is a short step from this to the techniques of grafting and budding. Even tissue culture, which must be done under very controlled, sterile conditions, is nothing more than an adaptation of natural cell division.

Propagation by seed is, despite its drawbacks, still a vital part of plant propagation. There are some plants, particularly annuals, where vegetative propagation is either impractical or impossible. Also, gardeners and commercial growers continually demand new, improved plants. While it is theoretically possible to produce completely new plants vegetatively (by genetic manipulation), most new introductions are the result of cross-pollination and hybridising. The resultant seedlings are rigorously assessed and the best of them are perpetuated by vegetative propagation.

How is it that plants are so easy to clone yet it is still extremely difficult to clone animals? Even the most advanced plant is a far simpler organism than an animal, but the real reason why asexual reproduction is so much easier in plants is almost entirely because of cells known as meristems.

Meristems are small clusters of cells found at the main growing points of the plant, pri-marily the root tips and the terminal leaf buds. Exactly what these *apical* meristems will develop into is largely determined by their position within the plant and the hormonal messages they receive. Generally, root-tip meristems develop into roots, while stem-tip meristems produce leaves, stems and flowers. However, meristematic cells also occur along the stems of plants, where their primary function is to increase the diameter of the stem as the plant grows. This thin *lateral* meristem layer is known as the vascular cambium or cambium layer and is vitally important in plant propagation because it has the ability to produce apical meristem cells, and it can fuse with the cambium of other plants.

If we take a section of stem and insert it into the soil, the vascular cambium may produce meristem cells that lead to root development. If it does, we have a successfully struck cutting. If we take a piece of stem from one plant and keep its cambium in close contact with that of another, they may

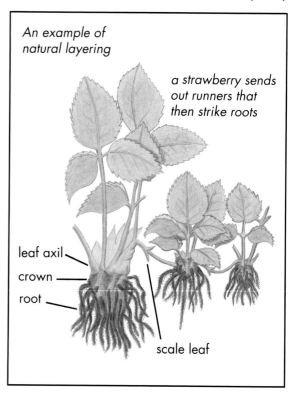

An example of natural layering

a strawberry sends out runners that then strike roots

leaf axil

crown

root

scale leaf

The true flowers of *Osteospermum* (syn. *Dimorphotheca*) 'Pink Starry Eyes'.

fuse. If they do, we have a successful graft.

Complex chemical changes are involved in these processes and the development of the growth phytohormones, known as cytokinins and auxins, that initiate root and stem development may take some time. Also, this development time varies between plant types. Consequently we get variations in strike rate and strike times between plants. Meristem tissue only occurs in the higher plants, so algae, fungi and ferns cannot be grafted or propagated by cuttings. However, such plants are often very easily divided.

This is a very simple explanation of why propagation works. However, it is not imperative that you understand the botany involved before becoming a successful plant propagator. What is important is that you understand why particular methods are used, when to use them and what to expect.

Which method?

Choose seed when diversity does not matter and vegetative means when you want exact copies of the parent plant.

There are two occasions when you would automatically propagate by seed: when growing annual bedding plants and vegetables, and when producing new hybrids or true species. Most garden annuals have been carefully inbred to come as true to type as possible from seed and they would be difficult to produce in the huge numbers required by any other means. True species usually reproduce so close to type that any variability is unlikely to be noticeable.

Hybridising between different plants, on the other hand, deliberately introduces diversity. Hybridisers select their parent plants for characteristics they would like to see in the offspring. Crossing the parents and raising the seed produces a range of seedlings, some of which may combine the best characteristics of both parents with none of their vices. Coming up with a plant that best encapsulates the ideals of the hybridiser often takes several generations of crosses and back breeding. Once bred, the new hybrid will then have to be propagated vegetatively to maintain its characteristics.

Most of the more permanent garden plants, the perennials, shrubs and trees, are hybrids or selected naturally occurring forms. These plants would not reproduce true to type from seed and must be propagated vegetatively. The same holds true for any unusual growth forms that may occur, like a changed flower colour or variegated foliage. If these new characteristics are to be preserved, then the plant must be propagated vegetatively. Also, hybrids are often infertile so there is no alternative but to propagate them asexually.

While there is only one method of sexual propagation, there are many ways to propagate a plant vegetatively. The simplest method is division, which is most commonly used with tubers and fibrous-rooted perennials. Cuttings are widely used for producing all sorts of plants: soft-wood cuttings for perennials and some shrubs; semi-ripe cuttings for most shrubs and conifers; and hardwood cuttings for hardy deciduous shrubs and trees. Bulbs and corms have their own special techniques, or they may be grown from the numerous natural offsets they produce. Grafting and budding are generally reserved for hardwood plants that are difficult to grow by other means or would not do well on their own roots. Occasionally, soft-stemmed plants like tomatoes and cucumbers are grafted onto super-vigorous rootstocks to produce strong-growing, heavy-cropping plants.

Some plants can only be propagated using one of the above methods, but most can be grown in several ways. The method you choose will largely depend on the type of plant, the propagation material available, the number of new plants you require and your level of experience. Plant propagation, especially on the domestic scale, is an inexact science where minute variations in propagation material, growing facilities and climate can make a vast difference to the results, so be prepared to try anything. If it works for you stick with it.

Timing

Unless you have a particular interest in a specialist area, such as roses or fruit trees, that requires specialised techniques, you will probably find that growing from seed, division and cuttings make up at least 95% of your propagation. These are simple techniques that are easily mastered, and often their success or failure depends not on how well they are executed but on *when* they are done. Timing is all important.

Plant growth is primarily controlled by a combination of temperature and day length, which is usually related to the prevailing season. Although under controlled conditions we can manipulate and regulate the way plants grow, most small-scale propagation is done under natural conditions.

The stage of growth that a plant has reached and the season often determine whether or not it is a good time to attempt to propagate it. This is not vitally important when sowing seed because it is easy enough to use a heating pad or even the heat from the hot water cylinder to provide enough warmth for germination. Likewise a refrigerator is a useful tool for simulating winter. Timing, however, is an important factor in deciding when to divide a plant, take a cutting, graft or bud. The precise time for each technique varies markedly depending on your local climate. In mild areas many evergreen shrubs can provide semi-ripe cuttings year round, but where winters are cold propagation may have to stop because the plants are completely dormant and the ground is frozen solid.

Throughout this book you will find guides

Cultivars, such as *Scabiosa* 'Blue Butterfly', *Veronica teucrium* 'Royal Blue' and *Campanula perscifolia* 'Telham Beauty', are produced by crossing different parent plants.

to the right times for the various propagation methods, but remember these are only an indication, it is impossible to be precise. For example, spring is a clearly defined time of year, but if a particular plant performs best when the days are over 12 hours long and the temperature above 18°C, the time when the best results will be achieved could vary by several weeks depending on your locali-ty, how controlled your propagation environment is and your level of experience.

Some people seem to know almost instinc-tively the right time to propagate a plant; most of us, however, need guidelines. The best guide is your experience: if it worked last year at this time, the chances are that it will work this year; if a certain seed germi-nated well in autumn, why sow it in spring? It is a good idea to keep a diary so that you know what worked and what didn't, and when. Nevertheless, don't let your experi-ence prevent you from listening to others or stand in the way of experimentation.

Chapter 2

TOOLS OF THE TRADE

A recently struck softwood cutting is lifted by a
plant label acting as a dibbler.

FOR me, the keys to successful plant propagation are a good working environment, simple, functional tools and an uninterrupted flow of work. Although very few tools are required for basic propagation, there is one you can't do without — curiosity. When you want to propagate a plant, find out everything you can about it; read the books, ask at the local nursery, take a horticultural course if need be. Also, take the time to learn the botanical names of your plants and try to become adept at plant identification. If you can identify a plant, a whole world of information becomes available to you.

Working environment

Your working environment should include somewhere with a bench at which you can stand or sit comfortably for long periods of time and somewhere to keep your plants. If you have a greenhouse they are usually one and the same place. However, a greenhouse is not essential. You can get by using your garage or tool shed bench to work on, with a home-built cold frame in which to keep the plants. A cold frame is a large box with a glazed or clear plastic-covered lid. Frames are usually built so that the lid slopes down from the back to the front to allow for easy access and so that the rain runs off. Because a frame would rapidly heat to very high temperatures if kept in the sun, a position in light shade or with early morning sun only is best.

A cold frame is effectively a miniature greenhouse, and it can be used to harden off tender seedlings prior to planting them out and as somewhere to keep trays of cuttings until they strike. Cold frames can be very efficient but they have one major drawback: being so small they are subject to wide variations of temperature — they heat up quickly and cool down just as fast. A covering of hessian (burlap) or frost-cloth will provide some additional insulation on cold winter nights but it is unlikely to stop hard frosts penetrating. Adding heating cables to a frame can eliminate this problem but the cables require an electricity supply. Consequently heated frames are not usually stand-alone structures.

With the introduction of cheap and efficient plastic films, frames have largely been replaced in commercial nurseries by plastic-skinned tunnel houses. These have the advantage of being large enough to work in while not costing much more than an extensive collection of frames. However, frames are reasonably efficient, cheap to construct and can be made portable, so they are still a good choice for home gardeners, especially where space is limited.

If space is really at a premium, you might consider a small self-enclosed propagating unit. These usually comprise a plastic tray with a fitted clear-plastic dome-like lid and a heated pad on which to place the tray. These have a place in small-scale production, or if you have some particularly fussy plants that are best kept on their own. But with so little room for the plants to develop, I think they are best restricted to seed raising only. Also, the heating pad tends to dry out the soil in the tray very quickly, which can be disastrous for young plants.

Be warned that plant propagation is an addictive hobby and eventually you may decide that you must have a greenhouse. The little 1.8 m by 2.4 m greenhouses commonly sold by the large hardware chains are

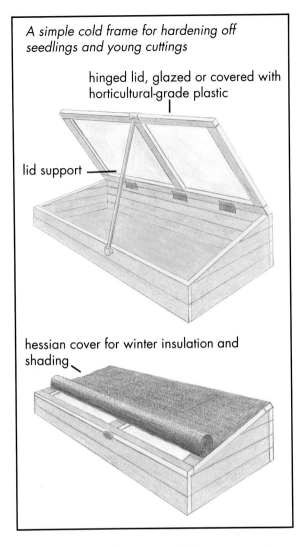

A simple cold frame for hardening off seedlings and young cuttings

hinged lid, glazed or covered with horticultural-grade plastic

lid support

hessian cover for winter insulation and shading

really too small. You will have very little room to move and the house will be subject to wide temperature variations. They are possibly suitable if you don't mind having your work bench somewhere else, but if you want to be able to work with your material at the propagation site a 2.5 m by 3 m greenhouse is the minimum practical size. Here you will have enough room for a small bench and about twenty propagation trays, with a little space left over to harden off a few plants. Whatever type of greenhouse you choose, it should be easy to extend because, inevitably, you will want more room.

There are several types of greenhouse construction and they vary considerably in cost, functionality and durability. The plastic-covered tunnel house design is the cheapest and easiest to erect, but some have sloping sides that intrude too far into the working space. Glasshouses are sturdy, long-lived and often simple to extend, but they are expensive and prone to breakage. Keen propagators usually end up with a glasshouse because once you have a favourite hobby it is easier to justify the expense and, in the long run, it really is the best option. However, if you're keen on do-it-yourself projects or really can't afford a glasshouse, the best compromise is a wooden frame with straight sides and a pitched roof covered with heavy duty (200μ) agricultural plastic film. Bear in mind that even the best modern plastic will only last about five years before it begins to deteriorate. Rigid polycarbonate or fibreglass sheeting lasts longer, although it can be more expensive than glass.

Summer ventilation is vitally important. A shadecloth cover will cool the greenhouse, but you must also provide adequate ventilation. The rule is to have an area of venting roughly equivalent to 30% of the floor area of the greenhouse, so a 2.5 m by 4 m (10 m²) greenhouse would require 3 m² of venting.

If you live in a frost-free area, your greenhouse may not require winter insulation, otherwise consider it essential. A single layer of plastic or glass will not stop a heavy frost damaging the plants. Lining the greenhouse with an additional layer of plastic will make a considerable difference, although the ultimate system is the purpose-built double-skinned or double-glazed greenhouse with two layers of glass or plastic separated by an air gap. This provides frost protection down to about -7°C for greenhouses over 75 m², or -3°C for those down to 5 m², without additional heating.

If you intend to grow only hardy plants, or live in a very mild climate, you may not need a greenhouse, but you will still need a sheltered environment to protect your tender young plants. A shadehouse protects plants from hot sun, reduces moisture stress, lessens wind velocity and also provides a little frost protection. Shadehouses also provide a good intermediate level of protection for plants that are being hardened off after being moved out of a greenhouse.

Modern shadehouses are usually simple wooden or pipe frames covered with shadecloth, which is a woven plastic fabric that is attached to wires strung from the frame. Shadecloth is available in various densities of weave, which is usually stated as a percentage: a 32% shadecloth lets through two-thirds of the light and stops one-third while a 50% shadecloth allows through about as much light as it stops. This is not a true measure of the shading effect of the cloth in relation to full sun, but it provides a guide. It is usually best to opt for the lightest cloth that provides enough protection or you may find that plants suffer from sunburn when moved out to the open garden.

Basic equipment

Your hardware needs depend on how many plants you want to propagate, what those plants are and how sophisticated and automated you want your propagation set-up to be. Elementary propagation, raising a few seeds and divisions, needs little more than some seed trays, potting mix and a sharp knife. However, you should consider the following as being necessary for a basic system that is capable of producing a wide range of plants.

• A *top-grade pair of secateurs* is your most important tool. Don't skimp on this item, good secateurs will last a lifetime. Choose an easily disassembled scissors-style with replaceable blades and make sure they are comfortable to hold and use.

• Two *sharp knives* are also necessary. First, a very sharp knife with a fine blade for

A home-built propagating house with white shadecloth sides and a plastic-film-covered roof.

budding, grafting and any other precise cutting work. Hobby knives with snap-off replaceable blades are cheaper that fancy propagating knives and always have a razor-sharp edge. Second, a general purpose knife for chores like cutting string, and for dividing small perennials and cutting tubers. Two hobby knives would be fine.

• A *good quality spade* for breaking up large clump-forming perennials, such as *Hosta* and *Phlox*. A steel-handled spade is preferable because it will be used for levering. Keep your spade sharp and dividing woody clumps will be quick and easy.

• A *sharpening stone* for keeping your secateurs, knives and spade in top condition. A simple double-sided carborundum stone of the type that can be bought from any hard-ware shop is perfect.

• A *hand mister* for keeping your cuttings moist while you take them. Any small atomiser bottle, such as an old window-cleaner bottle, will do.

• *Trays* are used for most propagating but *pots* are suitable if you have only a small number of cuttings, or are sowing very fine seed. The containers should have ample drainage holes and be easy to wash. You will need something to cover your container with: a pane of glass is ideal for trays, an upturned saucer for pots. Cuttings should be kept in a close humid environment. If you do not have a misting unit, enclosing the tray or pot in a tent made from a plastic bag is probably the easiest solution, but make sure you keep the tray out of direct sun or the delicate cuttings will cook very quickly.

You will also need a selection of small pots, sometimes called tubes, into which struck cuttings and seedlings can be potted.

• A *soil sieve*, either plastic- or wooden-framed with a 6-mm stainless steel mesh, is an essential item for making consistent cutting and seed-raising mixes.

• *Root-forming hormones* are produced by the plant but adding a little extra speeds things up. Root-forming hormones come in powder, liquid and gel forms. They are not essential, but can make a considerable difference with hard-to-strike cuttings.

• *Two thermometers* are useful to have. A soil thermometer for measuring the temperature of composting potting mix, heated beds and garden soil, and a maximum/minimum thermometer for keeping track of the conditions in your propagating area.

• *Insecticides, fungicides* and *soil sterilising* agents, either organic or synthetic, have a place in pest and disease control but careful management is more important.

Soil sterilisation is generally unnecessary provided you always use fresh potting mix. It's a false economy to enhance poor soil mixes with sterilising agents or bactericides and many of these compounds are extremely toxic. Steam is the only safe sterilant. Consign used mix to the compost heap or use it as a garden mulch.

I prefer bark-based potting mixes because they are fairly consistent and sterile (as far as diseases are concerned, not weeds). Peat or fern fibre mixes are good but they tend to be more variable. Avoid using soil-based potting mixes for propagation as they are an invitation to disease and will almost certainly need sterilising.

Insecticides are largely unnecessary if you select clean propagating material. Of course your greenhouse will need to be kept pest free, as will any plants that show signs of damage. *Bacillus thuringiensis* (a naturally occurring disease that attacks caterpillars), derris dust and pyrethrum-based sprays are safe and effective.

Cuttings and freshly germinated seedlings are often attacked by fungal diseases. Good ventilation is the most important preventative. If diseases do occur, sulphur and copper sprays are effective, but be careful as they can damage tender young growth.

• In addition to the above list you will need sundry equipment like *marker pens* and *labels, cleaning rags, tape, wire ties*, etc. It is also useful to keep a *diary* so that you know when certain jobs were done and for keeping weather records.

There are other aids that are useful but not essential. Undoubtedly the one that would make the greatest difference to your cutting production is an automated mist propagating unit. A mister will keep your cuttings moist and turgid (firm) even under the most trying conditions by ensuring that the cutting's foliage remains moist and never gets to the point of wilting.

These units work through a moisture-sensitive device located among the cuttings which opens a solenoid valve as it dries out. This valve controls nozzles that deliver a fine mist. Once the moisture level has been restored, the sensor switches off the water.

Most mist units are set up as permanent fixtures above a heated bed, but you can buy portable misters. They range from simple one tray affairs to all-in-one mist systems that take the form of a plastic tray, a little like a shower liner, with heating cables built into it along with mist nozzles and a moisture sensor. A misting unit may seem like an extravagance but it takes the guess-work out of cutting propagation, frees you from the labour of constantly tending softwood cuttings in warm weather, is a good way of gently moistening fine seed after sowing and can be useful in raising fern spores. Even the most sophisticated system will quickly pay for itself.

Fogging is a development of the misting principle. Instead of having a relatively coarse mist that is switched on and off as

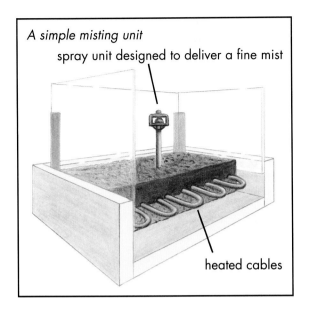

A simple misting unit

spray unit designed to deliver a fine mist

heated cables

needed, the plants are continually enveloped in a very fine cloud of fog. This keeps the air at maximum humidity without excessively wetting the soil.

You will need mains water pressure and an electricity supply if you intend to run mist or fog. However, modern plastic plumbing simplifies installation of the pipework and the electrical work is simple enough that you only really need an electrician to inspect your work and make the final connections.

Hygiene

Good hygiene is vitally important regardless of the scale of your propagation operation. Keep your tools and containers clean, use fresh potting mix for each new tray of seeds or cuttings and discard spent mix as soon as possible. You should always wash propagating trays and pots before use and after emptying them, preferably with a disinfectant,

but at least with a high pressure burst from the hose.

You should also regularly (every two months or so) clean your frames or greenhouse to keep down the incidence of pests and diseases. Household detergents and disinfectants (preferably biodegradable) work well but keep them off the plants. A thorough annual clean-up is also a good idea.

Keep an eye out for insect or disease trouble, and try to eliminate weeds. It is important to prevent young plants becoming smothered by weeds. These are jobs that are easily done provided you keep on top of them, but if the weeds and pests are allowed to proliferate you may have real problems.

Water

The quality of your water supply can have a marked effect on your plants. Excessive quantities of chlorine and other salts will eventually damage plants. If you know that your local water is chemically treated, it may be a wise move to contact your local authority to see if it has any known harmful effects, and if so, find out what corrections may be required.

Be wary of any obvious white deposits that form on the outside of propagating trays and pots, particularly those under mist. It is quite normal for a crust to form around the inside edge of the pot, this is simply a scum that develops from the minerals in the soil and can be washed off when the container is emptied, but deposits on the outside of the pots tend to indicate mineral build-ups, particularly calcium, in the water.

Chapter 3

PEST AND DISEASE CONTROL

Aphids

PROVIDED you always select strong healthy growth to propagate from and maintain hygienic growing practices, you should not have too much trouble with pests and diseases. As propagation involves keeping a close watch on your plants, if any problems do occur they can usually be dealt with before they get out of hand.

Pests

The most common insect pests are aphids, thrips and whiteflies. They are not that difficult to control, although they breed rapidly. The new pheromone traps and colour lures are very good at indicating that a problem exists, but unless it is a very small infestation

you will probably also need to use a mild insecticide. This is because the traps can only catch the winged adults whereas it is the sap-sucking larvae that cause most of the damage. The larvae can be controlled with fatty-acid or oil-based sprays — complete coverage of the undersurface of the foliage, where the larvae feed, is essential.

Mites are also major pests, and they are particularly quick to build up resistance to chemical controls. Oil sprays work by smothering the mites with a fine oil and suffocating them. They are effective, but good coverage is especially important as mites often lurk in inaccessible places.

Cutworms, beetle larvae, porina moths and

Juvenile thrips on a rhododendron.

Whiteflies and their larvae. Note the fungus growing on the larval secretions.

weevils are not usually present in sufficient numbers to cause significant damage. Cutworms are large nocturnal caterpillars that are the larvae of several species of moth. Some beetles, moths and weevils have subterranean larvae that feed on plant roots but unless their numbers are very concentrated, as sometimes happens if they occur in potted plants, they are unlikely to cause unsustainable damage. Bad infestations of all of these pests can be controlled with soil insecticides, such as dazomet, or by drenching the soil with a safe pyrethroid insecticide. Do not use liquid derris as it is inactivated by soil contact. Cutworms are easily controlled by hand at night.

Sciarid flies, the larvae of which live in the soil, are common in potting mixes and can become quite serious pests in greenhouses and other warm growing environments. Under normal conditions they have little or no effect on general plant growth, but if their larvae are present in large numbers among young seedlings they can cause considerable damage to the roots. Control is the same as for weevils and beetle larvae.

Mealy bugs are strange-looking insects covered in a fine white powder. They cause damage through sap-sucking and tissue rasping. Adults can be removed by hand, but the larvae live in the soil and are best controlled by drenching with a safe pyrethroid insecticide.

Scale insects are sap-sucking insects with a protective outer case. They cause debilitation and disfigurement and their secretions often lead to sooty mould. Oil sprays, which suffocate the insects, offer the best control.

Larger pests such as slugs and snails can cause considerable damage to young seedlings. Chemical baits are effective but they may also poison birds and other animals, so be careful how you use them. Removal by hand is time-consuming and can be ineffective, but some of the home-made traps often mentioned in organic gardening books are worth trying.

Caterpillars and slaters can be controlled as they appear and in most cases do not present much of a problem.

Mites on a rose leaf.

Juvenile scale insects on the underside of a leaf.

Mealy bugs.

Diseases

More young plants, especially seedlings, are damaged or killed by fungal diseases than any other cause. This is because they have very soft growth that is easily bruised, which allows an entry point for such diseases, and because the mild, humid propagating environment that is so ideal for young plants is also ideal for the spread of fungus.

There are two main forms of fungal disease: the surface moulds, such as botrytis and powdery mildew, which show up as greyish deposits on the foliage or stems, and the wilt diseases, such as phytophthora, rhizoctonia and damping off, which most commonly show up as the rotting of soft tissue. Wilt diseases usually affect the stems of seedlings or the bases of cuttings but may occasionally cause foliage or stem discolouration and die-back.

Under greenhouse conditions fungal problems can occur at any time; in the garden they are usually at their worst in cool, humid conditions in autumn. Good ventilation reduces their effect by lessening the build-up of spores. Greenhouses and frames should always be well ventilated, even in winter. Uncover trays of young seedlings and pot on struck cuttings as soon as possible. Do not crowd your seedlings or cuttings, as this will encourage the spread of disease. When striking cuttings under mist make sure that the soil is not constantly waterlogged. It should be just moist enough to keep the cuttings turgid, any more will lead to rotting.

Systemic fungicides are undoubtedly the most effective control in a moist propagating environment, but most are unacceptable to organic gardeners. Organic sprays, such as *Melia* and *Sambucus* extracts, are useful but are best used as preventatives. If an outbreak occurs, colloidal sulphur and copper-based fungicides are very effective while being environmentally safe, but they may damage soft growth, so use them with care.

Rhizoctonia and phytophthora wilt are soil-borne and control by any chemical means, synthetic or organic, is seldom effective. Good drainage prevents these diseases spreading, but it may also be a good idea to look for a new source of potting mix. As mentioned earlier, do not be tempted to re-

A rose leaf showing signs of powdery mildew.

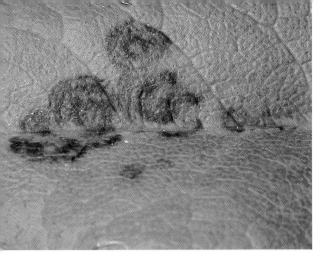

Necrotic ring spot virus.

A camellia with virus-induced variegation and some sunburn.

use your potting mix, unless it has first been thoroughly steam sterilised.

Keeping fungal diseases under control is largely a matter of good hygiene. Keep your propagating area well-ventilated, use fresh potting mixes and don't crowd your plants.

Occasionally you will also see viral diseases. These usually appear as a yellow mottling or flecking of the foliage, often with distorted new growth. Viruses are not always fatal but they are generally debilitating and nearly always incurable. Plants that are infected with viruses should be destroyed, preferably by burning. Do not put them in your compost and most definitely do not use them for propagating because any plants propagated from virus-infected stock will also be infected.

Careful use of sprays
Young plants with very soft growth (particularly young seedlings) may be damaged by some sprays, including organic sprays and strong oil sprays, so test any new spray on a few plants before introducing it for general use. The safest and most effective sprays are the fatty-acid-based insecticides and sulphur or copper-based fungicides.

You should aim to keep your use of sprays to a minimum, and one of the best ways of doing this is accurately timed preventative spraying. Rather than having to deal with fre-

quent outbreaks of pests throughout the growing season it is more efficient and effective to have your spray programme coinciding with crucial points in the pest's life cycle. For instance, many moths produce several generations of caterpillars in a normal growing season. Spraying at the egg laying to hatching stage for each of these generations will give you maximum control with a minimum of spraying.

Nobody wants to be exposed to pesticides and there are certainly bad environmental consequences caused by overuse. Synthetic agricultural chemicals are slowly getting safer but at the same time they are becoming less important as the newer organic controls, especially those based on naturally occurring diseases and pheromones, are now genuinely effective rather than the matter of faith they once were.

Because pests build up resistance to sprays, and because of our increased awareness of the dangers of some of the early chemicals, such as Lindane and DDT, new safer sprays are constantly being developed and introduced. However, many gardeners still have stocks of dangerous and outdated sprays. If you have such material, don't be tempted to use it, get rid of it. But do it responsibly; contact your local authority for their approved methods of disposal.

Chapter 4

Table 1: GENERAL PROPAGATION METHODS

Lupins and *Weigela*.

THE following table lists the main methods of propagation for over 1000 common genera. Note that individual species or cultivars may have special requirements. For example, many herbaceous genera contain both annual and perennial species, and the propagation techniques used for them may vary. Refer to the later tables (seeds, division and cuttings) for more details. For example, if a genus is listed as being propagated by seed and by cuttings, make sure you check both the seed and cutting lists for any specific requirements.

Certain conventions have been followed when choosing which methods to list.

• As all flowering plants may be propagated from seed, this method is only listed when it is one of the most practical methods for producing garden plants or when required for producing grafting or budding stocks. For example, magnolias may take many years to flower when raised from seed, but seedling magnolias can be used as grafting stocks, so seed is listed as a method for producing magnolias.

• Division is listed for plants that produce suckers, offsets and natural layers as well as those with clumps of divisible roots or tubers.

• Plants listed as being produced from bulbs are those that can most commonly be bought as loose bulbs in their dormant season or those that can be readily lifted and stored as dried bulbs, corms or tubers.

• Virtually all plants (other than annuals) may be propagated by layering, so this is not listed.

• Most plants that can be grafted may also be budded, so the two methods are not differentiated.

Table 1: General propagation methods

Table 1 General Methods

Plant	Seed	Divide	Cutting	Bud/Gft	Bulb
Abelia			•		
Abeliophyllum	•		•		
Abelmoschus (okra)	•				
Abies	•		•		
Abutilon	•		•		
Acacia	•		•		
Acaena	•	•	•		
Acanthus	•	•	•		
Acer	•		•	•	
Achillea	•	•			
Achimenes			•		•
Acidanthera syn Gladiolus					•
Ackama	•				
Acmena	•				
Aconitum	•	•			
Acorus		•			
Actinidia (kiwifruit)			•		
Adenandra			•		
Adiantum		•			
Adonis	•				
Aesculus	•				
Aethionema	•		•		
Agapanthus	•	•			
Agapetes			•		
Agastache (anise hyssop)	•	•	•		
Agathis	•				
Agave		•			
Ageratum	•				
Agonis	•				
Ailanthus	•	•	•	•	
Ajuga		•	•		
Akebia			•		
Alberta	•				
Albizia	•				
Alcea (hollyhock)	•				
Alchemilla	•	•			
Alectryon	•				
Allamanda			•		
Allium	•	•			•
Alnus	•		•		
Aloe		•	•	•	
Alonsoa	•		•		
Aloysia syn Lippia			•		
Alseuosmia	•		•		
Alstroemeria	•	•			

Plant	Seed	Divide	Cutting	Bud/Gft	Bulb
Alyssum	•		•		
Amaranthus	•				
Amaryllis					•
Amelanchier	•	•			
Ammi	•				
Ampelopsis		•	•		
Anacyclus	•		•		
Anagallis	•		•		
Anaphalis		•	•		
Anchusa	•	•			
Andromeda	•		•		
Androsace	•	•			
Anemone	•	•			
Anethum (dill)	•				
Angelica	•			•	
Angophora	•			•	
Anigozanthus		•			
Annona (cherimoya)	•			•	
Anomotheca					•
Anthemis	•	•	•		
Anthericum	•	•			
Antholyza		•			
Anthriscus (chervil)	•				
Antigonon	•				
Antirrhinum	•		•		
Apium (celery)	•				
Aptenia			•		
Aquilegia	•	•			
Arabis	•	•	•		
Araucaria	•				
Arbutus	•		•		
Archeria			•		
Arctostaphylos		•	•		
Arctotheca		•	•		
Arctotis	•	•	•		
Ardisia	•		•		
Arenaria	•	•	•		
Arisaema		•		•	
Arisarum (mouse plant)		•			
Aristea	•	•			
Aristolochia	•		•		
Aristotelia	•		•		
Armeria	•	•	•		
Armoracia (horseradish)	•	•			
Arnica	•	•			
Artemisia	•	•	•		
Arthropodium	•	•		•	•
Arum true forms		•			•
Aruncus	•	•			
Arundinaria (bamboo)	•	•			
Arundo	•	•			

Plant	Seed	Divide	Cutting	Bud/Gft	Bulb
Brachyglottis			• •		
Brachysema					
Brassica	•				
Bravoa					•
Briza	•	•	• •		
Brodiaea	•	•			
Bromeliads	•	•			
Browallia	•				
Brugmansia syn					
Datura			•	•	
Brunfelsia		•	•		
Brunnera		•			
Brunsvigia					
Buddleia	•		•		
Bulbinella	•				
Bupthalmum	•				
Butia	•				
Buxus			•		
Caesalpinia	•		•		
Calamintha (calamint)		•	•		
Calandrinia	•		•		
Calceolaria	•		•		
Calliandra					
Callicarpa			•		
Callistemon	•		•		
Callistephus					
Callitris					
Calluna			•		
Calocedrus syn Libocedrus		•			
Calochortus					•
Calothamnus					
Catha			•		
Calycanthus	•		•		
Calytrix					
Camassia		•			•
Camellia			•		
Campanula		•	•		
Campsis		•			
Canarina					
Canna		•			
Cantua					
Capsicum	•		•		
Ceratonia					
Cardamine		•	•		•
Cardiocrinum		•			
Cardoon	•				
Carex	•	•	•		
Carica (papaya)	•	•	•		
Carissa		•	•		
Carmichaelia	•		•		

Plant	Seed	Divide	Cutting	Bud/Gft	Bulb
Asarina synMaurandya	•				
Asclepias (swan plant)	•		•		
Asparagus	•				
Asphodeline		•			
Aspidistra	•	•	•		
Asplenium	•	•	•		
Astartea	•	•	•		
Astelia	•	•			
Aster	•	•	•		
Astilbe	•	•	•		
Astrantia	•	•	•		
Athyrium	•	•			
Aubrieta		•	•		
Aucuba	•		•		
Aulax	•		•		
Aurinia			•		
Avocado	•		•		
Azara	•		•		
Babiana					•
Backhousia	•		•		
Baeckia		•	•		
Bambusa (bamboo)		•	•		
Banksia	•				
Baptisia	•				
Bauera	•		•		
Bauhinia	•		•		
Beaufortia	•		•		
Beaumontia			•		
Begonia	•	•	•		
Beilschmedia					
Belamcanda		•			
Bellis	•	•	•		
Berberidopsis			•		
Berberis	•		•		
Bergenia		•			
Berzelia	•				
Beschorneria	•	•	•	•	
Beta (garden beets)	•				
Betula (birch)	•	•			
Bignonia	•				
Billardiera	•		•		
Billbergia		•			
Blechnum		•			
Bletilla		•			
Boltonia	•	•	•		
Bomarea		•	•		
Borago (borage)	•	•	•		
Boronia	•	•	•		
Bougainvillea			•		•
Bouvardia	•				
Brachychiton	•		•		
Brachycome	•		•		

Plant	Seed	Divide	Cutting	Bud/Gft	Bulb
Carpentaria	•		•		
Carpinus	•				
Carpodetus	•				
Carthamnus (safflower)	•				
Carum (caraway)	•				
Caryopteris	•				
Casimiroa (sapote)			•		
Cassia	•		•		
Cassinia			•		
Castanea (edible chestnut)	•		•		
Casuarina	•				
Catalpa			•		
Catananche		•	•		
Cattleya		•		•	
Cavendishia			•		
Ceanothus			•		
Cedrela					
Cedronella (balm of Gilead)	•		•		
Cedrus	•		•		
Celastrus			•		
Celmisia	•				
Celosia	•				
Celsia	•		•		
Celtis	•				
Centauria	•	•			
Cephalaria	•		•		
Cerastium		•	•		
Ceratopetalum	•	•			
Ceratostigma		•	•	•	
Cercidiphyllum	•				
Cercis	•		•		
Ceropegia	•		•		
Cestrum	•		•		
Chaenomeles			•		
Chamaecyparis	•		•		
Chamaedorea	•				
Chamaelaucium			•		
Chamaemelum (chamomile)			•		
Cheiranthus	•		•		
Childanthus	•				
Chimonanthus			•		
Chionanthus	•				
Chionodoxa		•		•	
Chlorophytum	•				
Choisya			•		
Chordospartium	•	•	•		
Chorizema	•		•		
Chrysanthemum	•		•		
Chrysocoma			•		

Plant	Seed	Divide	Cutting	Bud/Gft	Bulb
Cichorium (chicory, endive)	•				
Cimicifuga		•	•		
Cinnamomum	•		•		•
Cissus	•			•	
Citrullus (watermelon)					•
Citrus	•		•		
Cladanthus syn Anthemis	•				
Clarkia/ Godetia	•				
Clematis	•	•	•		
Cleome	•		•		
Clerodendrum	•		•		
Clethra					
Clianthus	•				
Clitoria	•		•		
Clivia		•			
Clytostoma			•		
Cobaea	•		•		
Colchicum					
Coleonema			•		
Coleus			•		
Congea	•		•		
Consolida (larkspur)	•				
Convallaria		•			
Convolvulus	•		•		
Cooperia					
Coprosma	•		•		
Corallospartium	•		•		
Cordyline	•		•		
Coreopsis	•	•	•		
Coriandrum	•		•		
Cornus	•		•		
Corokia	•		•		
Coronilla	•		•		
Correa	•		•		
Cortadeia	•	•	•		
Corydalis	•				
Corylopsis			•		
Corylus (hazelnut)	•		•		
Corynocarpus	•	•	•		
Cosmos	•		•		
Cotinus	•		•		
Cotoneaster	•		•		
Cotula	•	•	•		
Crassula			•		
Crataegus	•		•		
Crinodendron	•		•		
Crinum		•			•
Crocosmia		•	•		•

Top table

Bulb	Bud/Gft	Cutting	Divide	Seed	Plant
	•	•			Diosma
	•			•	Diospyros (persimmon)
		•			Dipidax
		•			Distictis syn
					Phaedranthus
		•		•	Dizygotheca
•		•	•	•	Dodecatheon
		•		•	Dodonea
		•	•	•	Dolichos
		•	•		Dombeya
		•	•		Doodia
		•	•	•	Doronicum
		•		•	Draba
		•			Dracaena
		•			Dracophyllum
•			•		Dracunculus
		•			Dregia
		•			Drimys
		•		•	Drosanthemum
		•		•	Dryandra
		•			Dryas
		•			Duranta
		•		•	Dysoxylum
		•		•	**Eccremocarpus**
		•	•		Echeveria
		•	•	•	Echinacea
		•	•	•	Echinops
		•			Echinopsis
		•		•	Echium
		•			Edgeworthia
		•			Elaeagnus
		•		•	Elaeocarpus
		•		•	Embothrium
		•			Enkianthus
				•	Entelea
		•		•	Epacris
		•	•		Epidendrum
			•		Epimedium
		•			Episcia
•			•	•	Eranthis
•			•		Eremurus
		•		•	Erica
		•	•	•	Erigeron
			•	•	Erinus
		•		•	Eriobotrya
		•			Eriostemon
		•	•	•	Erodium
				•	Eruca (sweet rocket)
		•		•	Erysimum
		•		•	Erythrina
•		•	•	•	Erythronium

Bottom table

Plant	Seed	Divide	Cutting	Bud/Gft	Bulb
Crocus					•
Crotalaria	•		•		
Crowea	•		•		
Cryptomeria	•		•		
Cucumis (cucumber, melon)	•				
Cuphea	•		•		
Cupressocyparis			•		
Cupressus	•		•		
Curcubita (pumpkin, squash)	•		•		
Cyathodes			•		
Cyclamen	•				•
Cymbidium		•			
Cymbopogon (lemon grass)		•			
Cynara (globe artichoke)	•	•			
Cynodon (Bermuda grass)		•			
Cynoglossum	•				
Cypella					•
Cyperus (papyrus)		•			
Cyphomandra	•		•		
Cypripedium		•			
Cyrtanthus		•			•
Cytisus	•		•		
Daboecia			•		
Dacrydium			•		
Dahlia		•	•		
Dais	•		•		
Dampiera		•	•		
Daphne	•		•		
Daucus	•				
Davallia		•			
Davidia	•		•		
Decaisnea	•		•		
Delphinium	•	•	•		
Desfontainea			•		
Deutzia	•		•		
Dianella	•	•			
Dianthus	•	•	•		
Diascia	•		•		
Diascorea (yam)		•			•
Dicentra	•	•	•		
Dichondra	•	•			
Dictamnus	•	•			
Dierama		•			
Dietes syn		•			
Moraea					
Digitalis	•	•	•		
Dimorphotheca	•		•		

Table 1: General propagation methods

Plant	Seed	Divide	Cutting	Bud/Gft	Bulb
Escallonia	•		•		
Eschscholzia	•				
Eucalyptus	•		•	•	
Eucharis					•
Eucomis	•	•			•
Eucryphia	•		•		
Eugenia	•		•		
Euonymus			•		
Eupatorium		•	•		
Euphorbia	•	•	•		
Euryops			•		
Eutaxia			•		
Exacum	•				
Exochorda			•		
Fabiana			•		
Fagus	•			•	
Fatshedera			•		
Fatsia	•		•		
Feijoa	•		•		
Felicia		•	•		
Ferraria					•
Ficus			•		
Filipendula	•	•			
Foeniculum (fennel)	•	•			
Forsythia			•	•	
Fothergilla			•		
Fragraria (strawberry)	•	•			
Francoa	•	•			
Fraxinus	•			•	
Freesia					•
Fremontodendron	•		•		
Fritillaria					•
Fuchsia			•		
Gaillardia	•	•			
Galanthus					•
Galega	•	•			
Galium (sweet woodruff)		•			
Galtonia					•
Gamolepsis			•		
Gardenia			•		
Garrya			•		
Gaultheria	•		•		
Gaura	•	•			
Gazania		•	•		
Geissorhiza					•
Gelsemium			•		
Geniostoma	•				
Genista	•		•		
Gentiana	•	•			
Geranium	•	•	•		

Plant	Seed	Divide	Cutting	Bud/Gft	Bulb
Gerbera	•	•			
Geum	•	•	•		
Ginkgo	•		•		•
Gladiolus					•
Glechoma		•	•		•
Gleditsia	•		•		
Globularia	•	•	•		
Gloriosa			•		•
Glycyrrhiza (liquorice)		•	•		
Gomphrena	•		•		
Goodia			•		•
Gordonia			•		
Grevillea	•		•		
Grewia			•		
Greyia	•		•		
Griselinea			•		
Gunnera	•	•			
Gypsophila	•	•	•		
Habranthus					•
Haemanthus					•
Hakea	•				
Halesia	•		•		
Halimiocistus			•		
Halimium syn			•		
Helianthemum	•		•		
Hamamelis	•		•	•	
Hardenbergia	•		•		
Harpephyllum	•				
Haworthia		•	•		
Hebe	•		•		
Hedera			•		
Hedycarya	•				
Helenium	•	•	•		
Helianthemum			•		
Helianthus	•	•	•		
Helichrysum	•	•	•		
Heliophila	•		•		
Heliopsis	•	•			
Heliotropium	•		•		
Helipterum	•				
Helleborus	•	•			
Hemerocallis		•			•
Hepatica	•	•			
Hermodactylus	•				
Herniara		•			
Herpolirion	•	•			
Hesperantha			•		
Hesperis	•				
Heterocentron syn			•	•	
Heeria			•		
Heuchera	•	•	•		
Hibbertia	•	•	•		

Top table

Plant	Seed	Divide	Cutting	Bud/Gft	Bulb
Kerria	•	•	•		
Knightia	•				
Kniphofia	•	•			
Kochia					
Koelreuteria	•		•		
Kolkwitzia	•		•		
Kunzea	•		•		
Laburnum	•		•		
Lachenalia	•		•		•
Lactuca (lettuce)	•				
Lagerstroemia			•		
Lagunaria	•				
Lambertia					
Lamium	•	•	•		
Lampranthus					
Lantana	•		•		
Lapeirousia					•
Larix (larch)	•				
Lathyrus (sweet pea)	•				
Laurelia					
Laurentia			•		
Laurus	•		•		
Lavandula			•		
Lavatera	•		•		
Leonotis					
Leontopodium (edelweiss)	•				
Leonurus					
Leptospermum	•		•		
Leschenaultia			•		
Leucadendron			•		
Leucocoryne					
Leucojum		•		•	
Leucopogon			•	•	
Leucospermum			•		
Leucothe		•	•		
Levisticum (lovage)			•		
Lewisia	•		•		
Liatris		•			
Libertia		•			
Libocedrus	•		•		
Ligularia			•		
Ligustrum (privet)			•		
Lilium					•
Limnanthes	•				
Limonium syn	•				
Statice					
Linaria	•				
Linnaea			•		
Linum	•		•		
Liquidambar	•			•	
Liriodendron	•			•	

Bottom table

Plant	Seed	Divide	Cutting	Bud/Gft	Bulb
Hibiscus	•	•	•		
Hippeastrum syn					
Amaryllis	•				•
Hoheria	•		•		
Holmskioldia			•		
Hosta	•	•			
Hovea	•				
Hovenia			•		
Hoya			•		
Humulus					
Hyacinthoides syn					
Scilla, Endymion					
Hyacinthus	•				•
Hydrangea			•		
Hymenanthera	•				
Hymenocallis syn				•	
Ismene					
Hymenosporum	•		•		
Hypericum			•		
Hypocalymma			•		
Hypoestes			•		
Hypolepis		•			
Hyssopus (hyssop)	•		•		
Iberis	•		•		
Idesia	•				
Ilex	•		•		
Impatiens	•		•		
Incarvillea	•	•			
Indigofera	•		•		
Inula	•	•			
Iochroma			•		
Ipheion					•
Ipomoea	•				
Iris		•			
Isoplexis	•				
Isopogon	•				
Ixerba					
Ixia			•		•
Ixiolirion					•
Jacaranda	•		•		
Jasione	•				
Jasminum	•		•		
Jovellana			•	•	
Juglans (walnut)	•			•	
Juniperus (juniper)	•		•		
Justicia syn	•		•		
Jacobinia					
Kadsura			•		
Kalanchoe	•		•		
Kalmia	•		•		
Kalmiopsis	•		•		
Kennedia	•		•		

Plant	Seed	Divide	Cutting	Bud/Gft	Bulb
Liriope	•	•	•		
Lisianthus	•		•		
Lithodora syn					
Lithospermum	•	•			
Lithops	•	•			
Littonia	•	•			
Lobelia					
Lobivia	•		•		
Lobularia syn					
Alyssum					
Lomatia			•		
Lonicera			•		
Lophomyrtus syn			•		
Myrtus			•		
Loropetalum	•		•		
Lotus	•				
Luculia	•				
Lunaria (honesty)	•	•			
Lupinus	•	•			
Lychnis	•	•			
Lycopersicon	•				
Lycoris					
Lysichiton					
Lysimachia		•	•		
Lythrum	•	•	•		
Macadamia					
Macfadyena syn					
Doxantha					
Macleaya	•	•	•		
Magnolia	•		•	•	
Mahonia	•		•		
Malcolmia	•		•		
Malus	•		•		
Malva (mallow)	•	•	•		
Malvaviscus			•		
Mammillaria	•		•		
Mandevilla			•		
Manettia	•		•		
Marianthus	•		•		
Marrubium (horehound)	•		•		
Matricaria	•	•			
Matthiola (stock)	•		•		
Maytenus	•		•		
Mazus		•			
Meconopsis	•		•		
Melaleuca	•		•		
Melia	•		•		
Melicope			•		
Melicytus	•		•		
Melissa (balm)	•	•	•		
Mentha (mint)	•	•	•		
Mertensia	•	•	•		

Plant	Seed	Divide	Cutting	Bud/Gft	Bulb
Meryta	•		•		
Mesembryanthemum (ice plant)	•		•		
Metasequoia	•		•		
Metrosideros	•		•		
Michaelia			•		
Micromyrtus			•		
Mimetes			•		
Mimulus		•	•		
Mina		•		•	•
Mirabilis			•		
Mitraria			•		
Moluccella	•	•			
Monarda (bergamot)	•		•		
Monstera			•		
Montbretia					•
Moraea					
Morus (mulberry)			•		
Moschosma			•		
Muehlenbeckia			•		
Murraya	•		•		
Musa (banana)		•	•		
Muscari	•	•	•		
Mussaenda			•		
Mutisia	•		•		
Myoporum		•	•		
Myosotis	•	•	•		
Myrrhis (sweet cicely)	•				
Myrsine			•		
Myrtus	•		•		
Myosotidium	•				
Nandina			•		
Narcissus (daffodil)	•				•
Nelumbo (sacred lotus)	•	•			
Nemesia	•	•			
Nemophila	•		•		
Nepeta	•				
Nephrolepsis	•				
Nerine	•			•	•
Nerium			•		
Nertera	•	•	•		
Nicotiana	•	•			
Nierembergia	•	•	•		
Nigella	•				
Nomocharis					
Nothofagus					
Notholirion	•			•	•
Notospartium					
Nymphaea	•	•	•		
Nyssa (tupelo)	•				
Ochna	•				
Ocimum (basil)	•				

Plant	Seed	Divide	Cutting	Bud/Gft	Bulb
Phormium (N.Z. Flax)	●	●	●		
Photinia	●		●		
Phygelius	●	●	●	●	
Phylica	●		●		
Phyllocladus	●		●		
Physalis	●	●			
Physostegia	●	●	●		
Picea	●		●	●	
Pieris	●		●		
Pimelia	●		●		
Pimpinella (anise)	●				
Pinus	●		●		
Pisonia	●		●		
Pittosporum	●	●	●		
Plagianthus	●		●		
Platanus	●		●		
Platycodon	●	●			
Plectranthus			●		
Pleione		●			
Plumbago	●		●		
Plumeria (frangipani)			●		
Podalyria	●		●		
Podocarpus	●		●		
Podolepsis	●				
Podranea			●		
Polemonium (Jacob's Ladder)	●	●			
Polianthes (tuberose)		●			●
Polygala			●		
Polygonatum	●	●			
Polygonum	●	●	●		
Polypodium		●			
Polystichum		●			
Pomaderris	●		●		
Pontederia		●			
Populus (poplar)	●		●		
Portulaca	●		●		
Portulacaria			●		
Potentilla	●	●	●		
Poterium (salad burnet)	●	●			
Pratia	●	●	●		
Primula	●	●			
Prostanthera			●		
Protea	●		●		
Prunus	●		●	●	
Pseudopanax	●		●		
Pseudowintera	●		●		
Psidium (guava)	●		●	●	
Psoralea	●		●		
Pteris		●			
Pterocarya	●	●	●		
Pterostylis		●			

Plant	Seed	Divide	Cutting	Bud/Gft	Bulb
Odontoglossum	●	●			
Odontospermum	●		●		
Oenothera (evening primrose)	●	●	●		
Oldenburgia	●				
Olea (olive)	●		●		
Olearia	●	●	●		
Omphalodes	●	●			
Ophiopogon	●	●			
Origanum (marjoram)	●	●	●		
Ornithogalum	●	●			●
Orphium	●		●		
Orthrosanthus	●	●			
Osmanthus			●		
Osmarea			●		
Osteospermum	●		●		
Ourisia	●	●			
Oxalis		●			●
Oxypetalum syn Tweedia	●		●		
Pachysandra		●	●		
Pachystegia	●		●		
Paeonia (peony)		●		●	
Pandorea	●		●		
Papaver (poppy)	●	●			
Paphiopedilum		●			
Parahebe			●		
Parrotia	●		●		
Parsonsia	●		●		
Parthenocissus	●		●		
Passiflora	●		●		
Pastinaca (parsnip)	●				
Paulownia	●		●		
Pelargonium	●		●		
Pellaea		●			
Penstemon	●	●	●		
Pentas	●		●		
Perilla	●				
Pernettya	●	●	●		
Perovskia			●		
Persoonia	●				
Petroselinum (parsley)	●				
Petunia	●		●		
Phacelia	●				
Phaenocoma	●		●		
Phaseolus (beans)	●				
Phebalium			●		
Philadelphus	●	●	●		
Philodendron			●		
Phlomis	●		●		
Phlox	●	●	●		
Phoenix	●	●	●		

Table 1: General propagation methods

Plant	Seed	Divide	Cutting	Bud/Gft	Bulb
Pterostyrax	•		•		
Pulmonaria	•	•	•		
Pulsatilla	•	•	•		
Punica	•		•		
Puya	•		•		
Pyracantha	•		•		
Pyrostegia			•		
Pyrus (pear)	•		•	•	
Quercus	•				
Quintinia	•		•		
Ranunculus	•	•			•
Raoulia		•	•		
Raphanus (radish)	•				
Raphiolepsis			•		
Regelia			•		
Rehmannia	•	•	•		
Reinwardtia	•		•		
Reseda (mignonette)	•				
Rhabdothamnus			•		
Rhamnus	•		•		
Rhododendron	•		•		
Rhodohypoxis		•			•
Rhoeo		•	•		
Rhopalostylis (nikau palm)	•				
Rhubarb		•			
Rhus	•		•	•	
Ribes	•		•		
Ricinus	•				
Robinia	•			•	
Rochea			•		
Rodgersia	•	•			
Romneya	•	•	•		
Romulea					•
Rondeletia			•		
Rosa (rose)	•		•	•	
Rosmarinus			•		
Rubia (madder)	•	•			
Rubus	•	•	•		
Rudbeckia	•	•	•		
Rumex (sorrel)	•	•			
Ruscus		•			
Russelia			•		
Ruta (rue)	•		•		
Sagina	•	•			
Salix (willow)	•		•		
Salpiglossis	•				
Salvia	•	•	•		
Sandersonia		•			•
Sanguinaria		•			
Sanguisorba syn Poterium	•	•			

Plant	Seed	Divide	Cutting	Bud/Gft	Bulb
Sanseveria		•	•		
Santolina	•		•		
Sapium	•		•		
Saponaria (soapwort)	•		•		•
Sarcococca			•		
Sasa (bamboo)		•			
Satureja (savoury)	•				
Sauromatum				•	•
Saxifraga	•	•			
Scabiosa	•	•			
Schefflera			•		
Schinus	•	•	•		
Schizanthus	•				
Schizostylis					
Schotia	•				
Sciadopitys			•		
Scilla	•	•			•
Scirpus	•	•			
Scutellaria	•	•			
Sedum	•	•	•		
Selago	•				
Sempervivum	•	•			
Senecio	•		•		
Sequoiadendron	•		•		
Serissa					
Serruria	•		•		
Sesamum (sesame)	•				
Shortia		•			
Sidalcea	•	•	•		
Silene	•				
Sinningia syn Gloxinia	•		•	•	
Sisyrinchium	•				
Skimmia	•		•		
Smilacina		•			
Solandra	•		•		
Solanum	•		•		
Solidago (goldenrod)		•			
Soleirolia (baby's tears)		•	•		
Sollya	•		•		
Sophora	•		•	•	
Sorbus	•	•	•		
Sparaxis	•	•			•
Sparmannia			•		
Spartium	•		•		
Spathodea	•		•		
Spiloxene					
Spinacia (spinach)	•				
Spiraea	•		•	•	
Sprekelia	•			•	
Stachys	•	•			
Stachyurus	•	•	•		

31

Table continues across two panels. Dots (•) indicate suitable propagation methods for each plant.

Plant	Seed	Divide	Cutting	Bud/Gft	Bulb
Tricyrtis	•	•			
Trigonella (fenugreek)	•	•			
Trillium		•			
Tristania			•	•	
Triteleia				•	
Tritonia					
Trollius	•	•			
Tropaeolum (nasturtium)	•	•			
Tsuga			•		
Tulbaghia		•		•	
Tulipa					•
Ugni syn					
Myrtus ugni			•	•	
Ulmus (elm)	•		•	•	
Urceolina		•	•		
Vaccinium	•		•		
Valeriana	•	•			
Vallota	•				•
Vancouveria		•			
Velthemia	•				•
Verbascum	•	•	•		
Verbena	•		•		
Veronica		•	•		
Vestia			•		
Viburnum	•		•		
Vigna syn	•				
Phaseolus	•				
Vinca	•		•		
Viola	•	•	•		
Virgilia	•		•		
Viscaria	•				
Vitex	•		•		
Vitis (grape)			•	•	
Wachendorfia		•	•		
Wahlenbergia	•		•	•	
Watsonia		•	•		
Weigela	•		•		
Weinmannia			•		
Westringia	•	•	•		
Wisteria	•		•		
Xeranthemum	•	•			
Yucca		•	•		
Zantedeschia	•	•			•
Zauschneria			•		
Zea (sweet corn)	•				
Zenobia			•		
Zingiber (ginger)		•			
Zinnia	•				

Plant	Seed	Divide	Cutting	Bud/Gft	Bulb
Staphylea	•		•		
Stauntonia	•		•		
Stenocarpus	•				
Stephanandra		•	•		
Stephanotis			•		
Sternbergia	•				•
Stewartia	•		•		
Stokesia	•	•	•		
Stranvaesia	•		•		
Strelitzia	•	•			
Streptanthera					•
Streptocarpus	•	•	•		
Streptosolen			•		
Styrax	•		•		
Sutherlandia	•				
Swainsona	•		•		
Symphoriocarpos		•	•		
Symphytum (comfrey)	•	•			
Syringa		•	•	•	
Syzygium	•				
Tagetes (marigold)	•				
Tamarix			•		
Tanacetum (tansy)	•	•			
Taxodium	•		•		
Taxus	•		•		
Tecoma			•		
Tecomanthe			•		
Tecomaria			•		
Tecophilea	•			•	
Tellima	•	•			
Telopea (waratah)	•		•		
Templetonia			•		
Ternstroemia			•		
Tetrapathea			•		
Teucrium	•		•		
Thalictrum	•	•			
Thryptomene			•		
Thuja	•		•		
Thujopsis			•		
Thunbergia	•		•		
Thymus (thyme)	•	•	•		
Tibouchina			•		
Tigridia	•				•
Tilia (lime)	•		•		
Tolmiea		•			
Torenia	•				
Toronia	•				
Trachelospermum			•		
Trachymene syn	•				
Didiscus	•				
Tradescantia		•	•		
Tragopogon (salsify)	•				

Chapter 5

GROWING FROM SEED

The berries of *Aucuba japonica* 'Crotonoides'

A SEED is an amazing thing; it contains all the genetic information necessary to produce an adult plant, be it a tiny alpine perennial or a giant forest tree. Seeds come in many shapes and sizes, ranging from the minute dust-like grains of the begonias through to giant palm seeds, yet they all share similar features.

Seed structure

All seeds are composed of an embryo with an endosperm (stored food) enclosed within a seed coat. Seed coats vary considerably, some are hard, black and shiny, others are beautifully patterned and interestingly textured, yet they all serve the same function —

to protect the embryo.

The endosperm often makes up the bulk of the seed. We are all familiar with the endosperm; it is the part of the seed we consume when we eat peas, beans and nuts. It is a concentrated store of the fats, carbohydrates and/or oils that are essential for keeping the developing embryo alive until it has functioning roots.

The embryo, which is usually very small in relation to the seed, is the young plant. The other parts of the seed are only there to protect or nourish the developing embryo. If the embryo is damaged the seed is useless.

The size of a seed is not directly related to the final size of the plant. While there are

Seeds come in a wide range of sizes and shapes, which affects how they are sown.

few small plants with large seeds, many of the forest giants have developed from something considerably smaller than your thumbnail.

Classification

The flowering plants (angiosperms) come in two forms, monocotyledons and dicotyledons. Monocotyledons, such as grasses and palms, produce only one seed leaf on germination. Dicotyledons, which are far more common among garden ornamentals, produce two seed leaves on germination. Dicotyledons include such diverse plants as beans, oaks and rhododendrons. The conifers (gymnosperms) are classified separately. Other important botanical differences between the groups include the way the seeds develop and how flower parts are grouped.

Collecting seed

The cheapest way to obtain seed is to collect it from your garden plants or those growing in the wild. As timing is important when collecting naturally occurring seed, plan ahead. When your chosen plant has finished flowering, look for swelling seed pods or ripening fruit. Determining when the seed is ripe is a matter of experience and judgement but there are generally visual clues. If there is any doubt, many seeds can be collected slightly unripe and will continue to ripen as long as they are in their seed pods.

When harvesting your own seed, it is important to remove as much of the chaff and surplus vegetable material as possible. If this is sown along with the seed it will tend to rot and may encourage fungal diseases. Once collected, the seed should be cleaned immediately. Dry seed pods and cones present no great problems. Hang them upside down in a paper bag in a dry, airy place and

The attractive rose hips of *Rosa virginiana plena*.

The fruit of the Himalayan strawberry tree (*Cornus capitata*).

wait for the seed to fall. All that is required from you is an occasional gentle tap to loosen them. A quicker method is to place the seed pods or cones in a saucer and break them up as they start to dry, thereby releasing the seed.

Extracting the seed from soft fruits can be awkward and messy. The easiest way is usually to crush the fruit, then steep it in water to remove the pulp. Tip off or lift out any floating material and the reasonably clean seed should be left at the bottom of the container. The fruit may need to be left in water for several days (this also softens the seed coat), and even then the seed may not come away completely clean.

Hybridising

Without doubt, there is nothing more satisfying than producing a new hybrid that is clearly an improvement on anything previously available, but it is not a quick process. Many hybridisers labour for years without producing anything that is truly outstanding. The first step in successful hybridising is usually to decide on a speciality and then stick with it. Very few of the really successful hybridists have been generalists.

The process of hybridisation is much the same for all plants, it is just a matter of transferring pollen from the pollen parent to the seed parent. The seed parent should be a flower just on the verge of opening. Avoid opened flowers as they may have already been pollinated. If the stigma (the sticky pollen-receiving receptacle at the end of the style) is not easily accessible, prepare the flower by carefully stripping away the petals

Pandorea pandorana has large ornamental seed pods.

The pollen sac on an azalea stamen.

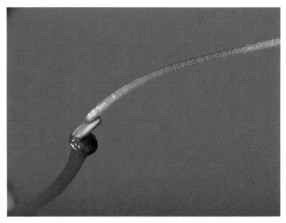

Azalea pollen transfer.

and removing the anthers to leave the stigma exposed.

Pollen can then be transferred by removing the anthers from the pollen parent and applying the pollen directly to the stigma of the seed parent, or by using a small paintbrush to pick up and transfer the pollen grains.

Once the pollinated seed has ripened, you collect and prepare it as described above. Some plants have more accessible flower parts than others, so practise with something simple, such as fuchsias or lilies, then decide on a speciality.

Hybridising has its difficulties. One of the most common problems occurs with double flowers. Unfortunately in producing the extra petals many or all of the stamens of a double flower may become modified into petals, thereby rendering the flower sterile, as it cannot produce pollen. Such flowers are only suitable as seed parents, although a very thorough search will occasionally reveal a functioning anther.

Another problem you may face is when you want to cross two plants that flower at different times. The only way round this is to preserve some pollen from your pollen parent until the potential seed parent is in flower. This is not difficult. You will need some gelatine medicine capsules, a desiccant (silica gel is the most widely available), a shallow jar or some other small container and access to a refrigerator.

When the pollen parent is in flower, remove a few fresh stamens and trim them down to just the anthers and a short length of filament so that they will fit in the medicine capsules. Line the base of your jar with a thin layer of desiccant and place the capsules on the desiccant. Place the jar, unsealed, in the refrigerator.

The gelatine capsules are semipermeable and the moisture from the stamens will be drawn out by the desiccant, which you should replace daily. After a few days the stamens will have become thoroughly dry. Once dried, refresh the desiccant and seal

A cutaway view of *Delphinium cardinale* showing the parts of the flower.

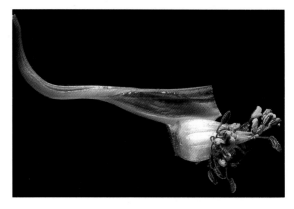

The flower structure of *Passiflora* 'Eynsford Gem' is very obvious, making it a good candidate for hybridising.

the container. The dried pollen can then be stored, usually indefinitely, in the freezer until needed. Allow the pollen to warm before use.

Buying seed

The most convenient way to obtain seed is to buy it and bought seed has several advantages over collected seed. Reputable seed companies will ensure that their seed is accurately named and true to type. The seed will have already been cleaned and will be ready for sowing or pre-sowing treatment. Also, seed packets often include valuable information about germination time, temperature, average germination percentage and the viability time. Buying seed may not be as satisfying as collecting and sowing your own, but it can take a lot of the guesswork out of the process and save you a lot of time.

Storing seed

Warmth and moisture are the greatest ene-

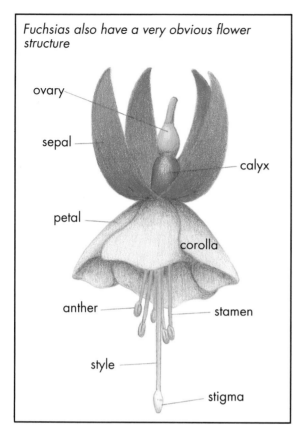

Fuchsias also have a very obvious flower structure

ovary

sepal

calyx

petal

corolla

anther

stamen

style

stigma

mies of stored seed. If you cannot sow seed immediately, it should be kept somewhere cool and dry. Commercial seed usually comes in vacuum-sealed packets with a best before date, but home gardeners rarely have the facility to vacuum seal their seed. The best alternative is to dust the seed with a fungicide and keep it in small paper bags in the bottom of your refrigerator. Paper bags, unlike plastic, allow the moisture to escape from the seed and the cool temperature of the refrigerator slows down the natural deterioration. You may find that very fleshy seeds with thin seed coats become desiccated under such conditions and these are best kept in plastic bags.

Variability

Plants grown from seed will always exhibit some variability and the greater the differences between the parent plants the more variable their progeny. However, extensive

breeding programmes, particularly those associated with flowering annuals and perennials, have brought us to the stage where we now have many very reliable seed strains.

Certain strains of perennials, shrubs and trees, especially those with simple foliage variations such as purple leaves, also reproduce reasonably true to type from seed. Nevertheless, unless you are content to raise species only, or wish to undertake your own hybridisation programme, propagation of plants by seed is best suited to the production of annual bedding plants and vegetables.

F1 hybrids

You will often find packets of seeds labelled as F1 (first filial) hybrids. These plants are the result of a cross between two species that always produce a consistent hybrid offspring. F1 hybrids are usually specially bred to produce strong vigorous plants with desirable characteristics.

There is no point in saving the seed of your superior performing F1 hybrids because they will not reproduce true to type. This is because F1 hybrids can only be produced by crossing the original parents. Each new batch of F1 seed is the result of repeating that cross.

Germination and viability

Many seeds require the existence of precise conditions before they will germinate. An understanding of the environment in which the parent plant naturally occurs will usually give you some idea of the conditions required and how long the seed is likely to remain viable. Seeds of plants that grow where rainfall is sporadic or virtually non-existent usually have an extremely long viability period, as the seed will often have to remain dormant for long periods until the next rain. Likewise the seed of plants from areas with long cold winters has to be able

The seeds of Iceland poppies (*Papaver*) germinate within 2–4 weeks at the right temperature. Above is the popular 'Windsor Mix'.

to survive the winter. Some seed actually requires to be exposed to the cold of winter before it will germinate.

How a plant disperses its seed may also have an influence on germination and viability. For example, many seeds are enclosed in soft bodied fruit, and may not germinate until their hard seed coating has been softened. In the wild this usually happens when the seed passes through the digestive system of an animal.

Breaking dormancy

Many seeds that are otherwise difficult to germinate become remarkably co-operative if you know how to break their dormancy. This is an area where good reference material and note-taking are of enormous help.

Commercial seed often has the information on the seed packet, but when in doubt don't be afraid to experiment. Usually the plant will provide you with clues. For example, fast-growing annuals will normally germinate rapidly, requiring nothing more than the right temperature and an initial watering to start them off, but plants that come from extreme climates may have particular requirements. Seeds from plants naturally occurring in very cold climates may require stratification; those from areas subject to prolonged drought may need soaking; and seeds with very hard seed coats may need scarification as well as soaking.

Stratification

Stratification, which is exposing the seed to cold, simulates natural winter chilling. Make up some seed-raising mix, moisten it, mix in the seeds and then put the seeds and soil in a plastic bag or, better still, a translucent plastic box with a lid. Leave the bag somewhere fairly warm, out of direct sun, for three or four days to allow the seeds to take up moisture and then transfer the bag to the refrigerator (not the freezer). Just how long the seed will need to be chilled depends on the species: 10 weeks is about the average.

Take the bag out each week and shake it to make sure the soil remains loose and aerated. If the seeds have started to germinate, it is definitely time to get them out of the refrigerator. If, after stratifying, the seed doesn't germinate, yet still appears quite sound, try a repeat period of chilling — some seeds have to stratified twice.

Scarification

Scarification is just a matter of abrading the seed coat in order to allow it to soften more quickly when exposed to moisture. To scarify small seeds, line a jar with coarse sandpaper, put the seed inside, screw on the lid, then shake them until the seed coat is well scratched. Larger or more stubborn seedcoats can be rubbed with a file or nicked with a sharp blade, but be careful not to damage the embryo.

Some seeds have a very hard seed coat that responds better to softening by soaking in hot water rather than by abrading. Other seeds, especially those of plants from areas with very irregular rainfall, contain germination inhibitors that prevent the seed from germinating until it is thoroughly moist. If these seeds were to germinate after a light shower they would be sure to perish, so they will not germinate until they have had a really thorough soaking.

To ensure the seeds are well softened and soaked use moderately warm, not boiling, water. Twelve hours' soaking at a temperature of around 45–50°C is usually adequate, after which they may have swelled slightly.

In extreme cases, seeds with particularly tough seed coats may require soaking in a mild acid solution to soften them. This simulates the effect of the seed passing through an animal's digestive tract. Use very dilute hydrochloric or phosphoric acid and extreme caution. This is definitely a last resort.

Germination temperature

All seeds have an optimum temperature range at which they germinate best, as well as thresholds above and below where they will not germinate at all. Fortunately, most seeds germinate over a wide range of temperatures. If you aim for somewhere between 15–25°C, with more tropical seeds tending to do better at the warmer end of the scale, you won't go far wrong.

If you only intend to germinate seeds in spring and summer, artificial heating may be unnecessary. However, if you wish to give your seeds an early start or keep them growing on well into winter, heated beds or heating pads will be essential. When using any sort of soil heating remember to make regular checks on the soil moisture as additional heat can quickly dry it out.

Seed-raising mixes

If you are only intending to sow one or two trays of seed then it is probably more convenient to use a seed-raising mix straight from the bag, but most commercial seed-raising mixes are less than ideal. You'll get far better results and a more consistent mix if you make your own.

By all means use one of the bagged mixes as your base, but find one that is bark, peat or fern fibre based, reasonably consistent and well composted. You don't want a mix that is too fresh, nor one that is poorly composted or showing signs of fungus or moulds. Once you have a good basic mix, you can adjust it to your own specifications.

I find the addition of finely chopped sphagnum moss, at the rate of 30–50% by volume, makes a world of difference to a seed-raising mix. Sphagnum contains natural fungicides and it retains huge amounts of moisture while remaining open and free

As it is a species, *Euphorbia veneta* will grow true to type from seed, unlike many hybrids.

from compaction; and it is easily penetrated by the fine roots of young seedlings.

The seed-raising mix should be light and airy, yet moisture retentive. The sphagnum will ensure that it is moisture retentive, but the mix may need improved aeration. In this case perlite or fine pumice should be added.

The final mix should be about one-third each of bark or peat based potting mix, fine sphagnum moss and perlite or pumice. Because the sphagnum is quite fibrous, the mix must be sieved or the roots of the young seedlings will get caught up in the strands of moss, which makes them difficult to prick out (pot on). Run the mix through a 6-mm mesh soil sieve. Do not wet it first or it will be next to impossible to sieve.

You can use substitutes for these ingredients and as you gain experience you will almost certainly develop your own mixes. As always, experiment, but avoid garden soil because it is not sterilised and will be full of weeds and diseases. Also, it will compact down like concrete. A little sand will open up the mix and improve the drainage, but too much and it becomes hard and caked and may contribute to waterlogging and drainage problems. Do not use vermiculite, which rapidly breaks down to a soggy mass, encourages algae growth and makes pricking out difficult.

Sowing seeds in containers

You can choose to sow your seed in containers or directly in the open ground. Which you choose is largely determined by the type of seed and its intended purpose. Most vegetable seed and many quick-growing annuals are sown directly where they will be grown. These plants germinate and grow so reliably that they can be left to get on with it.

If you want to sow seed out of season, you are unsure of how the seed will germinate

It is easier to see fine seed, and thus distribute it more evenly, if it is mixed with talcum powder.

A freshly sown tray of seed covered with glass and a newspaper.

and develop, or if the plants will be too tender to survive outdoors, then it is better to sow in containers. Container growing has the advantage of a more easily controlled environment, but there are disadvantages. Container-raised seeds will have to be pricked out at least once (every transplant is a shock and will check growth to at least a small degree) and they will require regular watering and occasional feeding.

Propagating trays are the best type of container for seed raising, although almost any shallow box or tray that has drainage holes will do. Fill the container to within a couple of centimetres of the top (this space allows the seedlings to develop before they hit the cover), then gently firm the mix into place and water with a fine mist. The tray is now ready for sowing.

Getting the seed evenly distributed over the tray is often the trickiest part of sowing. Very fine seed causes the most problems because it is hard to see. If you mix the seed with plain talcum powder or icing sugar you will be able to see the distribution pattern.

To cover or not to cover?

Beginners tend to bury their seeds. In most cases, all that is required is a very fine dusting of soil, if any at all. Some seeds need light to germinate and must not be covered. A good rule is to cover the seed to only just over its own depth, so fine seed is lightly covered while larger seeds are buried more deeply. Very fine seed, such as that of begonias and most ericaceous plants, should not be covered at all. Just keep it moist and it will germinate on the surface and send down roots.

Cover seed by lightly and evenly dusting it with finely sieved mix. Water it with a very fine mist or by soaking from below, then cover the tray with a pane of glass and place it in a warm place out of direct sun. Cover the glass with a sheet of newspaper to shade the seed; even seeds that need light for germination should be shaded or they may be cooked by the heat of the sun through the glass.

Aftercare

Seedlings are at their most vulnerable immediately after they have germinated. If they are too wet they may collapse due to fungal diseases and they are very easily destroyed by slugs, snails and birds. Uncover the seedlings by gradually raising the glass cover (prop it up with a stone, twig or piece of

Recently germinated seedlings.

Evergreen azalea seedlings at approximately three weeks.

wire) and then finally removing it. Good ventilation, combined with a mild fungicide, should control most damping off diseases.

Once your seedlings have their first true leaves, it will be time to transplant them from the seed tray into less crowded containers. If the young plants are very small, prick them out into a finely sieved 50/50 blend of potting mix and perlite, otherwise an ordinary potting mix is fine.

If for some reason you can't prick out the seedling immediately, it is essential that you feed them. Seed-raising mixes do not contain much fertiliser and rapidly growing young seedlings will quickly expend their reserves. A mild all-purpose liquid fertiliser will keep them happy for a while, but prick them out as early as possible.

Do not allow your seedlings to become smothered by weeds. Regular weeding eliminates competition, keeps the plants growing steadily and stops the weeds setting seed and compounding the problem. Remember to water the seedlings and feed them well and, most importantly, if they are naturally quick-growing plants do not allow them to become pot bound.

Sowing seed outside

When sowing in the open ground it is vitally important to prepare a good seed bed. While this is the key to getting good germination, the ultimate success of the crop also depends on good layout and planning. If you are planting out a vegetable garden, make sure that tall crops, such as sweet corn, will not be shading lower growers and that rampant growers, such as pumpkins, have adequate room to spread.

Prepare the seed bed by turning over the soil to ensure that it is loose and friable. This is also the time to add peat, compost and other soil conditioners. If you are adding compost, you may need to sieve it or break it up with a rotary hoe. The soil in the seed bed should be of a fine, even consistency and to ensure good root development there should be at least 15 cm of loose topsoil.

Young seedlings do not need a lot of feeding. However, as fertiliser is most easily applied before the plants appear, add a light dressing of general garden fertiliser and work it into the soil. If you have an acid soil, add a little lime. Dolomite lime (which has extra magnesium and other trace elements) is particularly good. Most vegetables, particularly brassicas, prefer a slightly alkaline soil, but over the years gardeners have tended to over-lime. A general dressing every two

Palm seeds, such as those of the Mediterranean fan palm (*Chamaerops humilis*), can be difficult to germinate and so are often sold pre-sprouted.

years is perfectly adequate and any specific plants that need lime can always be given extra. If you have any doubts about the nature of your soil, have it tested.

When you are ready to sow, use a string line to mark out the rows, then open a shallow trench (50 mm or so, depending on seed size) with the edge of a hoe. Sow the seed by shaking it carefully from the packet or by using a small hand sower (available from most garden centres). Aim for a close, even spread of seed along the row. Larger seeds are usually individually placed in position by hand. Very fine seed is normally sown quite heavily and thinned after germination; the thinnings can often be transplanted to make new rows. Once sown, gently close the earth back over the seed. Direct sown seed should be lightly covered or it may be moved around by the wind and rain.

Occasionally, you may find that the soil becomes caked and the seedlings have trouble pushing up through the surface crust. If this is a regular problem with your soil, it is a good idea to mix sand in with the soil you use to cover the seed. A couple of handfuls of coarse river sand every metre or so will keep the soil loose, and any crust that does form should break up quite easily.

After sowing, gently water the seed bed and keep it moist until there are signs of germination. Once the young plants are up, watch out for signs of fungal troubles, slugs and snails.

Pre-sprouting

Very large seeds, particularly monocotyledons (which produce only one seed leaf on germination) such as palms and gymnosperms (conifers), often present problems

Propagating ferns is an exciting process. Both *Cyathea dealbata* and *Dicksonia squarrosa* can be easily propagated from spores.

because they frequently send down a long tap root (radicle) that either goes through the drainage holes in the seed tray or is damaged by pricking out. Such seeds can be easier to handle if they are germinated before they are planted out. The easiest method of doing this is to soak the seeds for a few hours, place them on a tray lined with wet paper towels, cover them with more wet paper towels and put the tray in a warm dark place such as a cupboard near the hot water cylinder. Within a short time, the rapidly growing radicle should be apparent and the seeds can then be planted into individual pots of regular potting mix.

Some seeds, particularly palms and cycads, may be difficult to germinate or may take a considerable time. These plants are probably best left to the commercial seed houses that offer pre-sprouted seeds. There is usually a large price difference between ungerminated and pre-sprouted seeds, but it is a pretty fair bet that if pre-sprouting is offered then the seeds are difficult to germinate and the convenience of the pre-sprouted seed will be well worth the extra cost.

Propagating from spores

Spores, which are the way that many primitive plants, such as ferns, fungi and mosses, propagate themselves, are often looked on as a form of seed, although they are really quite different. Ferns, which are the most widely grown spore-bearing plants, can be propagated by vegetative means, but if large quantities are required and you have plenty of time, spores may be used.

Many ferns produce separate fertile and sterile fronds, the most common example of this habit is seen in the *Blechnum* ferns. The fertile fronds are very distinctive; they grow straight up from the crown and become covered in blackish brown sporangia or sori (spore capsules). As the sporangia mature, they rupture, releasing the spores. To collect the spores, hold a paper bag under a ripe fertile frond and tap the frond on the bag to release the spores. An alternative is to pick a few ripe fertile fronds, put them in the bag, then shake it.

Some ferns do not have distinct fertile fronds. In which case you will need to examine the undersides of the fronds to detect the presence of ripe sporangia.

Having collected your spores, the next step is to sow them. Fern spore does not remain viable for long and needs to be sown without delay. When the spores first start to grow, they produce a small liverwort-like scale called a prothallus. The prothallus contains both male and female genetic material. In order for the male and female cells to meet, the environment must be very moist, as the male gametes must actively swim or be carried in moisture to the female organ. So keeping the growing medium moist is essential when raising spores.

There are two common ways of sowing the spores. The first method is to sow them on a very finely sieved mix of sphagnum moss, peat and river sand, about one third of each by volume. Fill a pot with very wet mix, dust the spores over the surface, then gently water them and cover the pot with a pane of glass or plastic film. Place the pot in its saucer and fill the saucer with water.

Keeping the saucer topped up with water will ensure that the soil stays moist. Place the pot and saucer in a shady but reasonably warm position. The spores need light, but do not expose them to direct sun.

Unfortunately, potting mix nearly always contains spores of algae, liverworts and mosses, which will develop faster than the ferns and may smother them. For this reason I prefer a method that requires no potting mix and is clean and easy to use.

Find an old porous unglazed red clay brick. Scrub it clean using water and a very dilute bleach solution. Once clean, rinse it thoroughly to remove all traces of the bleach. This should ensure that the surface is sterile. Next, find a saucer or bowl large enough for the brick to sit in. Fill the saucer with water and place the brick in it, wide side down. After a few hours the top of the brick should become wet. If does not, you need a more porous brick or a deeper saucer.

When the brick stays wet, sow the spores on top. Cover the brick with a large plastic bag or some other clear cover and keep the saucer and brick in a cool place, making sure the saucer is constantly topped with water.

Within a few weeks a green film of prothalli will develop on the surface. Add a little liquid fertiliser to the saucer at this stage and, if the moisture level is right, small ferns will soon develop. When they are large enough to handle they may be gently prised from the brick and grown on.

Chapter 6

Table 2: GERMINATION REQUIREMENTS

Dianthus 'Caught in the Act'

THE following table lists the prime require-ments for success with all of the genera indi-cated as being suitable for raising from seed in Table 1. The temperatures indicated are the optimum for quick and even germination and the germination times apply only to seed sown at temperatures within the opti-mum range.

Pre-sowing treatment is optional in some cases and critical in others. It is wise to carry out the pre-sowing treatment even if it is only suggested.

The covering requirements have been abbreviated to three main types: 'Yes' if the seed must be covered and kept in a dark place until germination; 'Lightly' if just a thin covering of soil over the seed and light shad-ing of the tray is required; and 'No' if the seed needs light to germinate or is very fine.

Table 2 Seed Germination

Plant	Pre-Sowing	Temp	Cover	Germ
Abelmoschus (okra)	soak 12 hours	20-25	yes	10-15
Abutilon		18-25	yes	14-21
Acacia	scarify or soak 24 hrs	15-25	yes	10-35
Acanthus		18-22	yes	15-25
Acer	stratify 8-12 weeks	15-22	yes	14-42
Achillea		18-21	no	10-15
Achimenes		23-27	no	14-21
Acidanthera	soak 8 hours	18-24	yes	14-28
Ackama		15-22	lightly	21-42
Acmena		18-24	lightly	14-35
Aconitum	stratify 3 weeks	13-18	lightly	28-35
Aesculus	scarify & stratify 12 wks	15-22	yes	14-42
Aethionema		13-18	lightly	14-21
Agapanthus		21-25	yes	21-35
Agastache (anise hyssop)		15-22	lightly	7-21
Agave	soak 8 hours	18-24	yes	28-60
Ageratum		18-24	yes	18-35
Agonis		23-28	no	7-10
Ailanthus	better stratified 6 wks	18-24	lightly	14-35
Akebia	better stratified 4 wks	18-25	yes	21-42
Alberta	gently scarify or soak	15-22	yes	21-50+
Albizia	soak 24 hrs or scarify	18-24	lightly	21-56+
Alcea (hollyhock)		21-25	yes	15-30
Alchemilla		20-22	lightly	14-21
Alectryon	better soaked 8 hrs	13-20	lightly	10-28
Allium (chives)		18-24	yes	21-42
Allium (onions & leeks)		18-22	yes	10-15
Allium (ornamental)	stratify 4 weeks	18-22	lightly	10-14
Alnus	stratify 12 weeks	18-21	lightly	14-21
Aloe		15-22	lightly	18-35
Alonsoa		20-28	lightly	25-30
Alstroemeria		16-24	lightly	7-21
Alyssum		13-20	yes	15-50
Amaranthus		15-21	no	7-14
Amaryllis		20-25	lightly	8-12
Amelanchier	stratify 12 weeks	18-25	yes	14-35
Ammi		15-22	lightly	7-14
Anagallis		20-25	lightly	7-21
Anaphalis		18-24	lightly	5-10
Anchusa		18-20	lightly	7-10
Anemone		18-22	no	14-21
Anethum (dill)		15-21	lightly	14-21
Angelica	stratify 8 weeks	15-18	no	5-15
Angophora		15-21	yes	21-28
Anigozanthus		18-26	lightly	14-21
Anthemis		20-23	lightly	5-10
Anthericum		20-25	yes	14-28
Anthriscus (chervil)		15-21	lightly	7-14
Antigonon		21-25	yes	14-21
Antirrhinum		18-23	no	7-14
Apium (celery)	stratify 3 weeks	18-22	yes	10-20
Aquilegia		18-24	no	21-28
Arabis		18-23	no	7-15
Araucaria		18-25	yes	14-42+
Arbutus	remove all fruit pulp	15-22	no	14-35
Archeria		15-22	lightly	21-35
Arctotis		15-21	lightly	21-35
Arenaria		13-18	no	14-21
Arisaema	perhaps stratify 8 wks	15-22	yes	14-35
Armeria	soak 8 hours	18-21	lightly	7-14
Arnica		13-18	lightly	21-35
Artemisia (russian tarragon)				
Artemisia (wormwood)		15-21	lightly	18-28
Arthropodium		15-21	yes	7-14
Arum		15-22	yes	10-28
Aruncus	better stratified 6 wks	13-22	lightly	10-28
Asarina syn Maurandya		21-25	yes	14-35
Asclepias	better soaked 4 hrs	20-25	lightly	10-15
Asparagus	soak 8 hours	24-28	yes	10-28
Asphodeline		20-25	lightly	21-42
Astartea		18-25	lightly	28-35
Astelia		14-20	yes	14-35
Aster		18-21	no	14-35+
Astilbe		15-21	no	14-21
Astrantia		15-22	lightly	14-21
Aubrieta		18-21	no	10-28
Aulax	gently scarify or soak	18-25	yes	14-21
Avocado		20-27	partly	14-35+
Azara (vanilla tree)		15-22	lightly	21-50+
Babiana		15-24	lightly	14-35
Baeckia		18-24	lightly	10-28
Banksia	gently scarify or soak	18-24	yes	10-35
Baptisia	scarify	20-25	lightly	14-35+
Beaufortia		18-24	lightly	5-15
Begonia		20-25	no	14-28
Beilschmedia		22-26	lightly	10-35
Belamcanda		15-24	yes	14-21
Bellis		20-28	no	21-42
Berberidopsis		18-24	lightly	14-21
Berberis	stratify 8-12 weeks	15-22	lightly	7-15
Bergenia		15-22	no	14-35
Berzelia		18-24	lightly	14-35
Beta (garden beets)	soak 24 hours	20-24	lightly	15-20

Table 2: Germination requirements

Plant	Pre-Sowing	Temp	Cover	Germ
Betula (birch)	stratify if stored	13-22	lightly	10-28
Boltonia		15-22	lightly	10-21
Borago (borage)		18-21	yes	5-15
Bougainvillea		20-25	lightly	28-50
Brachychiton		20-27	lightly	18-35
Brachycome		18-21	no	7-10
Brassica crops		18-22	yes	7-14
Bravoa		18-24	lightly	10-35
Briza		18-22	lightly	5-10
Brodiaea		15-24	lightly	14-28
Browallia		21-25	no	7-15
Brunnera		20-28	lightly	14-21
Buddleia		20-25	yes	15-30
Bulbinella		15-24	yes	14-35
Bupthalmum		15-22	lightly	10-25
Butia	scarify & soak 48 hrs	20-27	yes	10-50+
Caesalpinia	soak 48 hours	20-25	yes	7-20
Calamintha (calamint)		14-22	lightly	10-21
Calandrinia	stratify 8 wks	14-22	lightly	10-28
Calceolaria		18-25	no	10-15
Calendula		18-23	yes	10-15
Calicarpa	stratify 8 weeks	15-22	yes	14-35
Callistemon		18-24	lightly	14-28
Callistephus		18-23	lightly	8-11
Callitris		18-24	lightly	14-35
Calluna	optional stratify	15-21	no	15-40
Calocedrus	stratify 12 weeks	14-22	lightly	14-35
Calochortus		15-22	yes	14-35
Calycanthus	better stratified 8 wks	15-22	yes	14-35
Camassia		12-20	lightly	14-28
Camellia	soak 24 hours	18-25	lightly	30-70
Campanula		18-22	lightly	14-28
Campsis	stratify 10 weeks	18-23	yes	14-21
Canna	scarify & soak 24 hrs	20-25	yes	7-14
Capsicum		20-24	yes	7-14
Cardamine	perhaps stratify 4 wks	12-18	lightly	8-15
Cardiocrinum	seed must be fresh	15-22	yes	14-35
Carica (papaya)	remove all fruit pulp	20-25	lightly	10-28
Carissa	remove all fruit pulp	18-25	yes	14-42+
Carmichaelia	scarify or soak 24 hrs	15-22	yes	21-100+
Carpinus	stratify 12 weeks	15-22	lightly	14-35
Carpodetus		15-22	lightly	10-35
Carthamnus (safflower)		18-22	no	5-15
Carum (caraway)		15-25	lightly	7-15
Cassia	scarify	20-28	yes	5-14
Castanea (edible chestnut)	scarify & soak 24 hrs	15-22	yes	14-35+
Catalpa	better stratified 6 wks	15-25	lightly	14-28
Catananche		18-23	lightly	5-15
Cedronella (balm of Gilead)		15-22	lightly	7-21
Cedrus	better stratified 8 wks	15-22	yes	14-42+
Celmisia		13-20	lightly	14-21
Celosia		22-25	yes	7-10
Celsia		20-25	lightly	10-15
Centaurea		18-21	lightly	7-14
Cerastium		16-21	no	7-14
Ceratonia	scarify & soak 24 hrs	18-25	yes	10-28
Ceratopetalum		18-24	lightly	10-28
Ceratostigma	better stratified 4 wks	15-22	lightly	14-35
Cercis	soak & stratify 12 wks	20-25	yes	20-90
Cestrum	soak at least 4 hours	18-24	yes	7-21
Chaenomeles	stratify 12 weeks	15-22	yes	14-35
Chamaedorea	seed must be fresh	20-28	yes	30-90
Chamaemelum (chamomile)		20-25	lightly	7-14
Cheiranthus		13-18	lightly	5-10
Chimonanthus	stratify 12 weeks	15-22	yes	14-3+
Chionanthus	seed must be fresh	15-22	lightly	14-35
Chordospartium	scarify or soak 24 hrs	15-22	yes	14-42
Chorizema	scarify or soak 24 hrs	18-24	yes	7-21
Chrysanthemum (annual)		15-20	lightly	10-14
Chrysanthemum (perennial)		18-22	lightly	7-28
Chrysanthemum maximum		18-22	no	10-14
Chrysanthemum parthenium		18-22	no	10-20
Cichorium (chicory/endive)		18-22	yes	7-14
Citrullus (watermelon)		20-24	yes	7-14
Citrus	better if seed peeled	18-24	yes	14-42
Cladanthus syn Anthemis		20-25	lightly	28-42
Clarkia/Godetia		16-22	lightly	5-15
Clematis	stratify 4 weeks	22-28	yes	30-100
Cleome		20-25	lightly	10-15
Clerodendrum		18-25	lightly	10-28
Clianthus	better soaked 8 hrs	15-24	yes	7-28
Clitoria	scarify & soak 24 hrs	20-25	yes	14-21
Clivia	seed must be fresh	22-26	yes	30-80
Cobaea		20-23	lightly	7-10
Coleus		20-25	no	10-15
Consolida (larkspur)	seed must be fresh	13-20	yes	10-21
Coprosma		15-22	lightly	10-28
Corallospartium	scarify or soak 24 hrs	15-24	yes	10-28
Cordyline		15-24	yes	14-35
Coreopsis (annual)		20-25	lightly	5-10

Plant	Pre-Sowing	Temp	Cover	Germ
Coreopsis (perennial)		16-22	no	10-15
Coriandrum (coriander)				
Cornus	stratify 12-20 weeks	15-18	lightly	7-14
Corokia		18-25	yes	30-300+
Coronilla	scarify or soak 8 hrs	13-22	lightly	14-28
Cortaderia		20-25	yes	15-40
Corylus	stratify 8 weeks	15-22	yes	5-10
Corynocarpus		20-25	yes	14-35
Cosmos		18-24	yes	14-21
Cotoneaster	stratify 8-12 weeks	15-22	yes	5-10
Cotula		15-22	yes	14-35
Crataegus	stratify 8 weeks	15-24	no	7-15
Crinum		15-22	lightly	14-35+
Cryptomeria	better stratified 8 wks	15-22	yes	14-35
Cucumis (cucumber)		20-24	yes	5-10
Cucumis (melon)		22-26	yes	7-14
Cuphea		18-22	yes	10-15
Cupressus	stratify 4 weeks	18-22	lightly	28-42
Curcubita (pumpkin/squash)		20-25	yes	5-10
Cyathodes		13-22	yes	14-42
Cyclamen (garden species)		14-18	lightly	21-28
Cyclamen persicum (pots)		14-18	yes	28-35
Cymbopogon (lemon grass)				
Cynara (artichoke)		20-25	no	5-15
Cynoglossum		20-24	lightly	15-25
Cypella		18-21	yes	5-10
Cyperus		18-24	lightly	10-21
Cyphomandra (tamarillo)		18-25	lightly	21-35
Cyrtanthus	seed must be fresh	20-25	lightly	14-28
Cytisus	soak 24 hours	18-24	yes	10-28
Daboecia		18-25	yes	21-35
Dacrydium	better stratified 8 wks	15-22	no	15-35
Dahlia		15-21	lightly	21-60+
Dais		20-25	yes	5-10
Daucus (carrot)		16-22	lightly	15-28
Davidia	stratify 12-24 weeks	15-22	lightly	10-15
Decaisnea	soak 8 hours	18-24	yes	14-21
Delphinium	seed must be fresh	15-24	yes	21-42+
Dianthus		18-22	yes	14-28
Diascia		15-22	lightly	12-18
Dicentra	stratify 6 weeks	15-22	yes	5-14
Dichondra		13-18	yes	7-14
Dictamnus	stratify 6 weeks	13-18	lightly	21-42

Plant	Pre-Sowing	Temp	Cover	Germ
Dierama		15-22	yes	14-35
Digitalis		15-28	no	5-15
Dimorphotheca		15-21	lightly	10-15
Diospyros (persimmon)	better stratified 8 wks	18-24	yes	14-35
Dizygotheca	stratify 4 weeks	20-26	no	35-40
Dodecatheon		15-22	lightly	28-42
Dodonea		15-25	lightly	145-28
Dombeya		20-25	yes	14-21
Doronicum		18-22	no	14-21
Dorotheanthus (Livingstone daisy)		17-25	lightly	14-21
Draba		13-18	no	10-28
Dracophyllum		13-22	lightly	21-42
Dryandra	gently scarify or soak	18-25	yes	14-35+
Dryas		15-22	lightly	42-70
Dysoxylum		18-24	no	14-28
Echeveria		20-27	no	14-35
Echinacea		20-25	lightly	5-15
Echinops	stratify 12 weeks	18-24	lightly	14-21
Echium		18-22	no	7-21
Elaeocarpus		15-22	lightly	21-42
Elaeagnus		20-25	yes	28-42
Entelea		18-25	lightly	10-20
Epacris		15-22	lightly	21-35
Eremurus		13-22	yes	14-35
Erica		15-22	no	15-35
Erigeron		13-18	lightly	14-21
Erinus		18-25	lightly	14-28
Eriobotrya	soak 24 hours	18-24	yes	10-28
Erodium		15-24	lightly	12-28
Eruca (sweet rocket)		15-21	lightly	5-10
Eryngium		18-25	lightly	5-10
Erysimum		13-18	lightly	5-15
Erythrina	scarify & soak 24 hrs	20+25	yes	7-21
Erythronium	better stratified 8 wks	13-20	lightly	14-35
Eschscholzia		18-22	lightly	5-10
Eucalyptus		18-25	lightly	14-28
Eucomis		20-25	yes	14-28
Eugenia		18-24	lightly	14-28
Eupatorium		15-22	lightly	10-28
Euphorbia (garden forms)	perhaps stratify 4 wks	18-22	lightly	5-15
Exacum		20-25	no	14-21
Fagus (grafting stock)	stratify 12 weeks	15-22	yes	14-35
Fatsia		18-25	lightly	28-40
Feijoa	remove all fruit pulp	18-24	lightly	10-28
Felicia	perhaps stratify 3 wks	14-21	lightly	21-35
Ficus		20-28	no	14-21
Filipendula		15-22	lightly	14-35

Table 2: Germination requirements

Plant	Pre-Sowing	Temp	Cover	Germ
Foeniculum (fennel)		18-22	yes	7-14
Fragraria (strawberry)	stratify 12 weeks	16-20	lightly	21-28
Fraxinus (ash)	soak 12 hours	15-22	lightly	14-35
Freesia		18-22	yes	21-28
Fremontodendron	scarify or soak 24 hrs	18-24	yes	21-42+
Fritillaria	stratify 8 weeks	15-21	lightly	14-35
Fuchsia		20-25	no	21-35
Gaillardia		20-24	no	5-15
Galanthus	better stratified 6 wks	13-20	yes	14-42
Galega		15-22	lightly	7-21
Galtonia		18-22	yes	14-21
Gardenia		20-25	yes	21-35
Gaultheria		13-22	no	21-56+
Gazania		18-24	yes	7-14
Geissorhiza		15-24	lightly	14-28
Geniostoma	stratify 4-6 weeks	15-24	lightly	14-35
Gentiana		18-25	lightly	14-28
Geranium		15-22	lightly	14-42
Gerbera		20-25	yes	10-15
Geum		18-22	no	8-20
Ginkgo	perhaps stratify 4 wks	20-25	yes	28-35
Gladiolus	stratify 8-12 weeks	18-25	yes	21-42
Gleditsia	scarify & soak 24 hrs	18-24	yes	7-21
Globularia	perhaps stratify 4 wks	13-18	lightly	7-14
Gloriosa		20-25	yes	28-35
Gomphrena		20-25	lightly	10-15
Goodia	scarify & soak 24 hrs	18-24	yes	7-21
Gordonia		15-22	yes	14-35
Grevillea	soak 12 hours	20-28	yes	14-35
Gypsophila		20-25	no	5-15
Haemanthus		18-25	lightly	14-42
Hakea	scarify & soak 12 hrs	18-25	yes	14-28
Halesia	better stratified 8 wks	15-22	lightly	10-28
Hamamelis	stratify 12 weeks	15-22	lightly	21-42+
Hardenbergia	soak 8 hours	18-24	yes	5-21
Harpephyllum	remove all fruit pulp	18-24	lightly	10-28
Hebe		13-24	yes	14-28
Hedycarya		15-22	lightly	14-35
Helenium		20-22	no	7-14
Helianthemum		20-25	lightly	14-21
Helianthus		18-24	lightly	5-10
Helichrysum		20-24	no	7-14
Heliophila		15-22	lightly	10-21
Heliopsis		18-22	lightly	5-15
Heliotropium		18-24	lightly	10-21
Helipterum		20-24	lightly	5-15
Helleborus	stratify 12 wks or sow outdoors	18-25	yes	30-300+
Hemerocallis	stratify 6 wks	15-22	yes	21-56
Hermodactylus	better stratified 4 wks	15-22	lightly	10-28

Plant	Pre-Sowing	Temp	Cover	Germ
Herpolirion		14-20	lightly	10-28
Hesperis		21-25	no	5-10
Heuchera		15-21	no	20-30
Hibiscus	soak 12 hours	20-25	yes	10-15
Hippeastrum syn Amaryllis		20-25	lightly	28-42
Hoheria		15-22	lightly	14-35
Hosta	perhaps stratify 6 wks	18-22	ligtly	14-21
Hoya		20-25	lightly	8-21
Humulus (hop)		20-25	lightly	21-35
Hymenanthera		15-22	lightly	14-35
Hymenosporum		18-25	yes	14-35
Hypericum		15-21	yes	10-21
Hypoestes		20-25	lightly	7-14
Hyssopus (hyssop)		15-21	lightly	7-14
Iberis annual (candytuft)		18-24	lightly	7-15
Iberis perennial	stratify 6 weeks	15-18	no	14-21
Idesia (wonder tree)	stratify 12 weeks	15-22	lightly	10-28
Ilex (holly)		18-22	yes	30-180+
Impatiens balsamina (balsam)		18-22	lightly	7-11
Impatiens walleriana		22-25	lightly	10-18
Incarvillea		15-18	lightly	21-35
Indigofera	scarify & soak 24 hrs	18-24	yes	10-28
Ipomoea	scarify or soak 12 hrs	18-24	yes	5-10
Iris	stratify 6-8 weeks	15-24	lightly	21-50
Isopogon	gently scarify	18-24	lightly	21-42+
Ixia		18-24	lightly	14-35
Ixiolirion		18-24	lightly	14-35
Jacaranda	better soaked 8 hours	18-28	yes	7-15
Jasione		18-22	lightly	10-15
Jasminum		20-25	lightly	14-35
Jovellana		13-22	lightly	7-21
Juglans	stratify 12 weeks	15-22	yes	14-35+
Justicia		18-22	no	14-35
Kalanchoe		18-24	lightly	10-15
Kalmia	stratify 12 weeks	18-22	lightly	21-42
Kalmiopsis		15-22	lightly	21-35
Kennedia	scarify & soak 24 hrs	14-20	no	5-21
Knightia	gently scarify	18-24	yes	21-42+
Kniphofia		15-24	lightly	21-28
Kochia		18-24	no	10-15
Kunzea		20-25	no	10-35
Laburnum	scarify or soak 24 hrs	15-24	lightly	10-35
Lachenalia		15-21	yes	28-70
Lactuca (lettuce)		18-24	lightly	14-28
Lagerstroemia		18-22	lightly	21-28
Lantana		20-25	yes	14-18
Lapeirousia	better stratified 8 wks	15-24	yes	35-60
Larix (larch)		15-22	lightly	14-35

Plant	Pre-Sowing	Temp	Cover	Germ
Lathyrus (perennial)	scarify or soak 24 hrs	13-18	yes	10-21
Lathyrus (sweet pea)	scarify or soak 24 hrs	15-20	yes	7-15
Laurelia		15-22	lightly	21-42
Lavandula (lavender)	stratify 4 weeks	18-24	no	14-21
Lavatera		18-22	yes	5-10
Leontopodium (edelweiss)	perhaps stratify 4 wks	18-22	no	15-21
Leonurus		18-24	lightly	10-28
Leptospermum		15-24	lightly	10-35
Leucadendron	gently scarify or soak	18-24	yes	21-42+
Leucopogon		13-20	lghtly	21-60+
Leucospermum	gently scarify or soak	18-24	yes	21-42+
Leucothe		15-22	no	14-35+
Levisticum (lovage)	stratify 4 weeks	15-21	lightly	7-14
Lewisia		18-22	lightly	28-35
Liatris		18-21	no	21-28
Libocedrus	perhaps stratify 4 wks	13-20	yes	21-42+
Lilium	some need stratifying	18-22	yes	21-42
Limnanthes		15-22	lightly	7-21
Limonium (statice)		18-22	yes	5-14
Linaria		13-15	lightly	10-15
Linum	stratify 8-12 weeks	18-22	lightly	14-28
Liquidambar	stratify 8-12 weeks	15-22	yes	14-35
Liriodendron	soak 24 hours	15-22	lightly	21-42+
Liriope		16-22	yes	28-35
Lisianthus		15-22	no	10-35
Lithops		20-25	lightly	10-15
Littonia		18-24	yes	14-35
Lobelia (annual)	stratify 12 weeks	20-25	lightly	14-21
Lobelia (perennial)		18-22	lightly	14-21
Lobularia (annual)				
Alyssum		18-25	no	7-10
Lunaria (honesty)	scarify or soak 24 hrs	18-24	no	10-15
Lupinus	better stratified 6 wks	18-24	lightly	5-15
Lychnis		18-22	lightly	21-28
Lycopersicon (tomato)		21-25	yes	7-14
Lycoris		15-22	lightly	10-28
Lysichiton	scarify & soak 48 hrs	13-20	lightly	14-35
Lythrum		18-22	lightly	14-21
Macadamia	stratify 12-16 weeks	18-25	yes	14-35+
Macropiper	better stratified 8 wks	18-25	yes	14-28
Magnolia	stratify 12 weeks	18-22	yes	30-90
Mahonia		15-22	yes	21-42
Malcolmia		14-20	lightly	7-21
Malus		15-22	yes	14-35
Malva (mallow)		18-22	yes	5-15
Marrubium		18-22	yes	10-15

Plant	Pre-Sowing	Temp	Cover	Germ
Matricaria		13-18	lightly	7-14
Matthiola (stock)		18-24	no	7-14
Meconopsis		16-22	lightly	14-28
Melaleuca		18-24	lightly	14-35
Melicytus		15-22	lightly	14-28
Melissa (balm)		18-22	no	10-15
Mentha		18-22	lightly	10-15
Meryta		18-25	yes	21-42
Mesembryanthemum				
Metasequoia	stratify 8-12 weeks	18-22	no	7-15
Metrosideros		15-22	yes	21-42+
Mimulus		15-25	lightly	14-35
Mina	scarify or soak 24 hrs	15-21	no	5-10
Mirabilis		18-22	yes	14-28
Moluccella		18-22	yes	5-10
Monarda (bergamot)		15-24	no	12-21
Monstera		15-22	lightly	14-21
Muehlenbeckia		24-28	lightly	14-21
Musa	remove all fruit pulp	13-20	lightly	10-28
Myoporum		20-27	lightly	10-28
Myosotidium		13-22	lightly	10-35
Myosotis		13-20	yes	14-28
Myrtus		18-22	no	7-14
Nemesia		15-25	lightly	14-35
Nemophila		16-22	yes	8-14
Nepeta (catmint/catnip)		16-22	yes	10-15
Nerine				
Nertera		18-22	yes	5-15
Nicotiana		15-22	lightly	14-42
Nierembergia		15-21	yes	14-28
Nigella		21-24	no	10-18
Nomocharis		20-25	lightly	5-15
Nothofagus	better stratified 8 wks	18-22	lightly	7-15
Notospartium	scarify or soak 24 hrs	15-22	lightly	14-35
Nyssa (tupelo)	stratify 8-12 weeks	15-22	lightly	14-42
Ochna	better soaked 24 hours	18-25	yes	10-28
Ocimum (basil)		18-24	yes	21-42+
Oenothera		20-25	lightly	7-21
Oldenburgia		18-24	no	5-10
Olearia	seed must be fresh	15-22	lightly	7-18
Omphalodes		13-22	lightly	10-28
Ophiopogon		15-22	lightly	14-42
Origanum (marjoram /oregano)		18-22	lightly	10-20
Ornithogalum		15-24	lightly	10-28
Orphium		15-24	lightly	5-10
Orthrosanthus	soak & stratify 4 wks	18-24	lightly	14-35
Ourisia		13-22	lightly	10-21
Oxylobium	scarify & soak 24 hrs	18-24	yes	7-21

Table 2: Germination requirements

Plant	Pre-Sowing	Temp	Cover	Germ
Pachystegia				
Paeonia	sprout then stratify 8 wks	13-20	lightly	14-28
Papaver (poppy)		18-22	yes	30-60
Papaver orientale		18-24	no	5-15
Parahebe		16-22	no	10-15
Parrotia		13-22	no	7-28
Parsonsia	stratify 8 weeks	15-22	lightly	14-35+
Passiflora		15-25	yes	14-42
Pastinaca (parsnip)	better soaked 24 hrs	20-25	lightly	28-45
Paulownia	perhaps stratify 2 wks	15-22	lightly	14-35
Pelargonium	scarify	18-24	yes	10-35+
Penstemon		18-25	yes	10-28
Pentas		13-20	yes	14-35
Perilla		20-24	no	5-15
Persoonia	gently scarify	18-22	lightly	7-21
Petroselinum (parsley)	better soaked 12 hrs	18-24	lightly	14-35+
Petunia		18-22	yes	14-21
Phacelia		22-27	no	7-14
Phebalium		13-22	lightly	7-14
Phaseolus (beans)		18-22	yes	14-35
Phlomis		15-22	lightly	14-21
Phlox (perennial)	stratify 4 weeks	13-22	yes	21-35
Phlox drummondii		18-22	yes	10-15
Phoenix	scarify & soak 8 hrs	15-18	yes	10-35
Phormium		18-25	yes	10-35
Phygelius		20-25	lightly	10-15
Phylica	remove all chaff	16-24	lightly	10-28
Phyllocladus		13-20	yes	21-60+
Physalis		15-21	no	7-14
Physostegia	stratify 12 weeks	18-21	lightly	7-14
Picea		13-20	yes	21-42
Pieris		15-22	no	15-35
Pimelia		15-22	lightly	14-28
Pimpiella (anise)		18-22	lightly	10-15
Pinus	some need stratifying	15-22	yes	14-42
Pisonia		15-25	yes	14-35
Pisum (peas)		15-22	yes	7-14
Pittosporum		15-22	yes	14-42+
Plagianthus		13-22	lightly	14-42
Platycodon		15-21	no	7-14
Plumbago		20-25	lightly	21-35
Podalyria	soak 24 hours	18-24	yes	14-35
Podocarpus	better stratified 8 wks	15-22	yes	21-60+
Podolepsis		16-22	lightly	10-28
Polemonium (Jacob's ladder)		20-25	lightly	18-28
Polygonum		20-25	lightly	18-28
Pomaderris		15-25	lightly	14-28
Portulaca		22-28	no	7-14
Potentilla	some need stratifying	15-22	lightly	14-21

Plant	Pre-Sowing	Temp	Cover	Germ
Poterium (salad burnet)		18-22	lightly	7-14
Primula	stratify alpines	15-22	no	14-28
Protea	gently scarify or soak	18-24	yes	15-42
Prunus	stratify 12 weeks	15-24	yes	21-42+
Pseudopanax	seed must be fresh	15-22	yes	14-35
Pseudowintera		13-22	yes	14-42
Psidium (guava)	better soaked 8 hrs	20-25	lightly	10-28
Pterostyrax	remove all fruit pulp	18-24	yes	14-35
Pulmonaria	better stratified 8 wks	13-22	lightly	10-28
Pulsatilla		13-20	lightly	14-21
Punica		20-25	no	20-28
Puya		13-22	lightly	14-35
Pyrus	better stratified 4 wks	15-22	yes	14-35
Quercus	stratify 12 weeks	15-22	yes	15-42
Quintinia	stratify 12 weeks	15-22	lightly	14-35
Ranunculus		15-18	lightly	14-21
Raphanus (radish)		13-21	lightly	3-7
Rehmannia		20-25	lightly	14-21
Reseda (mignonette)		20-25	no	5-10
Rhabdothamnus		18-25	no	14-35
Rhododendron		15-22	lightly	21-60
Rhoeo		18-24	yes	7-21
Rhopalostylis	soak 24 hours	18-25	yes	21-60+
Rhus	stratify 8 weeks	15-22	lightly	21-42+
Ricinus	better if soaked 8 hrs	20-25	yes	14-21
Robinia	better stratified 8 wks	15-22	yes	14-35
Rodgersia	better stratified 6 wks	13-20	lightly	10-28
Romulea		18-24	yes	21-42+
Rosa	stratify 8-12 weeks	13-22	yes	21-28
Rosmarinus (rosemary)		18-22	no	10-21
Rubia (madder)		14-20	yes	14-28
Rudbeckia		20-24	lightly	5-10
Rumex (sorrel)		13-20	lightly	7-14
Ruscus	stratify 12 weeks	15-25	yes	21-42+
Ruta (rue)		15-22	lightly	7-14
Sagina		13-18	lightly	10-15
Salpiglossis		20-24	lightly	14-21
Salvia		18-24	lightly	5-15
Salvia officianalis (culinary)				
Sandersonia	keep seed mix moist	18-22	yes	5-15
Santolina		18-24	lightly	21-60+
Saponaria (soapwort)		16-22	lightly	14-21
Sarcococca	better stratified 8 wks	18-22	yes	5-10
Satureja (savoury)		15-22	yes	21-42+
Saxifraga		18-22	no	10-20
Scabiosa		16-25	lightly	14-21
		18-22	no	10-20

Plant	Pre-Sowing	Temp	Cover	Germ
Schefflera		20-25	yes	14-28
Schinus		18-25	lightly	14-35
Schizanthus		15-22	no	7-14
Schotia	soak 24 hours	18-24	yes	10-28
Sciadopitys	stratify 8 weeks	15-24	yes	14-42
Scutellaria		15-24	lightly	14-35
Sedum		20-28	no	5-15
Sempervivum		23-28	lightly	14-35
Senecio (florist's cineraria)		21-24	no	10-15
Senecio (shrubby)		15-25	lightly	10-15
Sequoiadendron	better stratified 8 wks	15-22	yes	21-42+
Serissa		18-25	lightly	10-21
Serruria		18-24	yes	14-42
Sesamum		18-24	lightly	4-10
Silene		18-22	lightly	14-21
Sinningia syn Gloxinia		18-23	no	14-21
Sisyrinchium		16-24	yes	10-28
Solanum		18-22	no	7-14
Solanum (eggplant)		20-24	lightly	7-14
Solidago	better stratified 8 wks	13-20	lightly	14-28
Sophora	scarify or soak 48 hrs	15-25	yes	14-42
Sorbus (rowan)	stratify 12 weeks	13-22	lightly	18-35
Sparaxis	seed must be fresh	15-24	lightly	14-35
Spartium	scarify or soak 24 hrs	15-22	yes	7-21
Spinacia (spinach)	perhaps stratify 2 wks	10-15	lightly	5-15
Stachys		18-22	no	15-42+
Stauntonia	remove all fruit pulp	15-22	yes	15-42+
Stenocarpus	gently scarify or soak	18-26	yes	14-35+
Stephanotis		20-25	lightly	28-42
Stokesia		18-22	lightly	21-35
Strelitzia	soak up to 72 hrs	26-32	yes	30-180+
Streptocarpus		22-26	no	14-21
Styrax	perhaps stratify 6 wks	18-25	yes	14-35
Sutherlandia	soak 24 hours	18-24	yes	7-21
Swainsona	soak 24 hours	18-24	yes	10-28
Syzygium		15-24	lightly	14-35
Tagetes (marigold)		20-25	lightly	5-10
Tanacetum	stratify 12 weeks	15-22	lightly	7-15
Taxodium	stratify 12 weeks	18-22	yes	21-42
Taxus		18-25	yes	30-250+
Tecomanthe		18-25	yes	14-35
Tecophilea		18-24	yes	14-35
Telopea (waratah)	gently scarify or soak	18-24	yes	21-35
Templetonia	soak for 24 hours	18-24	yes	7-21
Tetrapathea	remove all fruit pulp	18-22	lightly	10-28
Teucrium		18-22	lightly	21-35
Thalictrum		18-22	lightly	14-35
Thryptomene		15-24	lightly	14-35
Thunbergia		21-24	lightly	7-14
Thymus (thyme)		18-22	yes	5-15
Tigridia		18-25	yes	18-28
Tilia (lime/linden)	stratify 12 wks	14-22	yes	14-35
Torenia		20-24	no	7-15
Toronia		18-25	lightly	14-35
Trachymene syn Didiscus		18-22	yes	14-21
Tradescantia		18-22	lightly	21-35
Trigonella (fenugreek)		14-22	lightly	4-10
Trillium	stratify 10 wks (twice)	15-21	yes	30-150+
Tropaeolum (nasturtium)		18-22	yes	10-15
Tweedia syn Oxypetalum	stratify 12 weeks	18-22	yes	10-15
Ulmus (elm)		15-22	lightly	18-42
Valeriana syn Heliotrope		18-22	lightly	21-28
Vallota		15-24	lightly	14-35
Veltheimia		18-22	lightly	10-28
Venidium		15-24	lightly	10-21
Verbascum		18-24	no	5-15
Verbena	seed must be fresh	15-22	no	5-15
Veronica		20-24	yes	7-14
Viburnum	stratify 10 wks (twice)	21-25	yes	42-90+
Viguiera		18-24	lightly	10-28
Vinca		18-22	yes	14-21
Viola		15-24	lightly	7-15
Virgilia	soak 24 hours	20-25	yes	10-28
Viscaria		18-24	yes	7-14
Vitex		15-25	lightly	14-42
Wachendorfia		15-22	yes	14-42
Weinmannia		18-25	lightly	21-42
Wisteria	scarify & soak 24 hrs	18-25	lightly	28-42
Xeranthemum		15-22	yes	10-15
Xeronema		13-18	lightly	14-35
Yucca		18-22	yes	18-28
Zantedeschia		16-25	no	28-42
Zea (sweet corn)		20-25	yes	5-10
Zelkova	stratify 12 weeks	20-24	yes	21-42+
Zephyranthes		14-22	lightly	14-42
Zinnia		20-24	lightly	5-10

Chapter 7

PROPAGATION BY DIVISION

Dicentra spectabilis

DIVISION involves little more than breaking up established clumps into a number of smaller pieces. Provided the clump can be broken up so that each division has a growth point and a few roots, the divisions should be capable of surviving as independent plants. Most perennials fall into this category and are by far the most commonly divided plants.

Division is a method that is well suited to garden production because it requires no specialised equipment and produces large, near mature plants. However, many perennials that can be divided will also grow from cuttings. *Delphinium* and *Dicentra*, for example, will strike readily from cuttings of their early spring shoots. Growing from cuttings is preferable when you want to produce large numbers quickly or when you don't want to disturb an established rootstock.

Rosettes, runners and offsets

The easiest plants to divide are those that form clumps of foliage rosettes. These plants, such as *Ajuga* and the saxifrages, break up neatly into new plants, each with a rosette of foliage and some roots. This simple separation can, in most cases, be done at any time of year.

Delphiniums will strike readily from cuttings of their early spring shoots.

Saxifrages break up neatly into new plants, each with a rosette of foliage and some roots.

Runners are side shoots that spread along the soil surface. In many cases small plantlets will grow from these runners. The plantlets will tend to strike roots wherever they come in contact with the soil so that in time they can be removed from the parent plant and treated as individuals.

Many cacti and rosette-forming succulents produce side shoots or new rosettes, known as offsets, that separate from the parent plant by themselves. All you need to do to produce a new plant is to pot them up. You can speed up the process by removing the offsets before they separate. They will have no roots at this stage, but pot them up anyway and the roots will soon develop. It is often a good idea to leave potting very succulent offsets for a few hours after removing them from the parent plant so that the damaged tissue has a chance to dry and form a callus.

Herbaceous perennials

Deciduous herbaceous perennials, such as *Hosta*, *Lythrum* and *Phlox*, are best divided near the end of their dormant period or just as they are starting into growth. The growth points are easier to identify at this stage and the plants can begin to grow as soon as they

Aloe saponaria can be broken up into rosettes at any time of year.

The growth habits of *Astelia* and *Phormium* (above left) and thymes (above right) allow for easy division.

have been divided. Getting the plants growing and the wounds healed as quickly as possible is the best way to avoid the soft rots that may occur on the cut surfaces. Dusting the wounds with a fungicide before planting will also help.

Evergreen and semi-deciduous herbaceous perennials often have distinct foliage clusters and heavy fibrous-rooted crowns. Many plants of the lily family, such as flax (*Phormium*) and *Libertia*, fall into this category. They too are best divided in late winter or early spring, but they may need to be planted out into nursery beds until established because they often have very few roots per division.

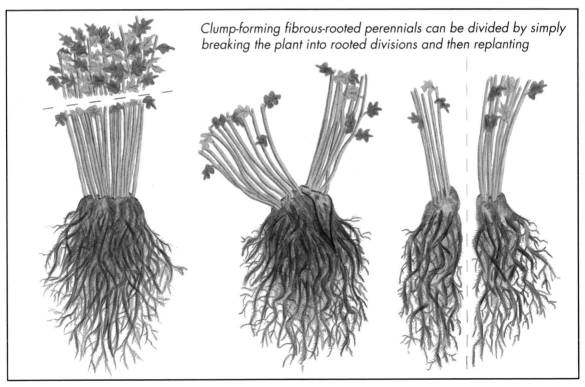

Clump-forming fibrous-rooted perennials can be divided by simply breaking the plant into rooted divisions and then replanting

Natural layering and aerial roots

Natural layers may form when a stem is kept in contact with the soil for a prolonged period. Many perennials and shrubs, particularly ground covers such as thyme (*Thymus*), regularly form natural layers that can be removed and grown on.

Some plants, usually tropical in origin or native to very humid areas, produce aerial roots that grow directly from their stems, often some distance above ground. A few climbers, such as ivy (*Hedera*) and *Campsis*, have aerial roots that are modified into grasping tools. At each node a tight mass of roots form and force themselves into any nook or cranny, thus supporting the plant as it climbs. If the stem is cut off below the aerial roots and planted, the roots will often develop as normal subterranean roots. However, this is not always successful and is best done under greenhouse conditions. Treat the piece as you would a cutting without roots until it starts to make new growth.

Ivy (*Hedera* spp. covering bridge at left) has aerial roots that when planted often develop into normal subterranean roots.

Suckers

Suckers are strong-growing shoots that emerge from the base of some trees and shrubs. Plants that produce suckers include lilacs (*Syringa*), elms (*Ulmus*), toon trees (*Cedrela*) and flowering quince (*Chaenomeles*). Often, the suckers can be removed with roots attached and grown on. Suckers of deciduous plants should be removed as early as possible in the growing season to ensure that they have adequate time to become established before they lose their foliage.

Rhizomes, tubers, bulbs and corms

Rhizomes, tubers, bulbs and corms are modified stems that are used as food storage organs. Plants that produce these organs are usually capable of withstanding an extended period of dormancy, during which they sur-

Syringa species often form suckers. If they have roots attached they can be grown on.

Peonies grow from crown rhizomes, but can be treated like other clump-forming perennials.

vive by using their stored energy as they cannot produce food by photosynthesis. Rhizomes, tubers and corms are capable of being divided and some bulbs can be separated into scales that can be grown on.

Rhizomes

Rhizomes grow on or just below the soil surface and often look like a plant stem. They are composed of segmented stems with buds at the nodes just like a stem above ground. The difference is that at each of the nodes there will also usually be some roots.

As each node has both a leaf bud and a root bud, propagation is simply a matter of breaking up the rhizome at the nodes. Mints and the creeping grasses, such as the infamous kikuyu and couch grass, are common rhizomatous plants.

Breaking up rhizomes is a simple matter, but the divisions are often lacking in true roots and foliage. For this reason, it is better to err on the generous side with your divisions. Dust any cut surface with a fungicide – powdered sulphur is suitable. Water and feed fresh plants well once they are growing. Like succulents, very fleshy rhizomes are better left to dry a little after breaking up to lessen the risk of rotting.

Some herbaceous perennials, such as asparagus and peonies, grow from what is known as a crown rhizome. As far as propagation is concerned these plants can be divided just like any other clump-forming herbaceous perennial rather than being treated like rhizomes.

Tubers

Tubers are always found underground at the base of the main stems of the plant. They are usually swollen and are easily removed from the fibrous root system. The tubers form as the plant grows. At the end of the season the top growth dies away leaving the tubers underground. They, in turn, sprout and grow to repeat the cycle the following season. The most common tuber is undoubtedly the potato, and it is typical of the general style of growth.

Plants that form tubers are very easily propagated. At the end of the season the crown is lifted and the tubers separated from it. They usually break away without any difficulty, but if they are stubborn, let them dry a little, then carefully prise them apart. If you have very well-drained soil the tubers may be planted out right away, otherwise store them in barely moist sawdust in a frost-free place and plant them out in spring.

Most tubers have several growth points or

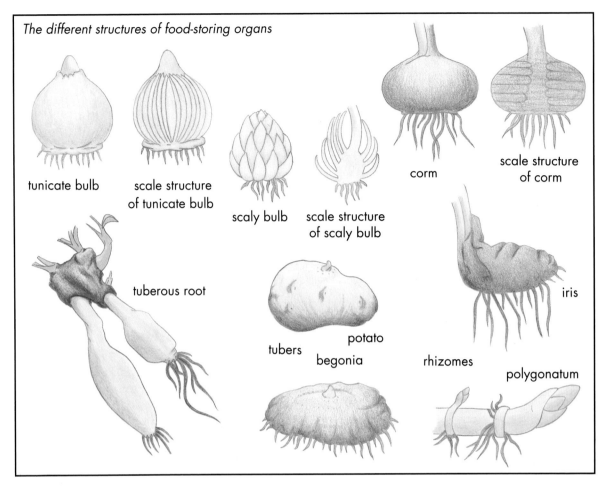

The different structures of food-storing organs

tunicate bulb

scale structure of tunicate bulb

scaly bulb

scale structure of scaly bulb

corm

scale structure of corm

tuberous root

tubers

potato

begonia

rhizomes

iris

polygonatum

eyes. For maximum production the tubers may be cut into smaller pieces, each with an eye. The best time to cut up tubers is in early spring, just as the eyes are starting to swell. There is always the possibility that an eye may be damaged or may fail to develop, so it is best to keep to at least two eyes per division.

Each division will have a large area of exposed flesh, which will make it very prone to fungal diseases and rotting. Dip the cut sections of tuber in a fungicide for protection — flowers of sulphur is safe and easy to use. Unless the tuber is quite shrivelled, do not plant the pieces right away, instead leave them in a warm dry place out of direct sun for a day or so while the cut surface dries and calluses form.

It probably goes without saying, but when planting out make sure that the eyes are facing up — the tuber's, not yours!

Corms

Corms and bulbs look similar to one another but there are significant differences that determine the methods of propagation suitable for each type. Bulbs are composed of layers of tissue called scales and they increase in girth each year by adding more layers of scales. Corms produce a completely new structure each year; the old corm withers away while a new and larger corm forms on top. Often small corms, or cormels, form between the old and new corms. These are quite viable but usually take several years to bloom. Some mature corms, such as those of *Crocosmia*, live for more than one season and may eventually become very large.

Fritillaria meleagris (above left) stores its food in a bulb, while *Anemone nemerosa* (above right) has a tuber.

Corms usually have a papery covering which can be peeled away to reveal the fleshy structure beneath. Doing so will also reveal that the upper surface of a corm has growth eyes similar to those of a tuber. Corms may be cut up in the same manner, leaving at least one eye per section. This is done in spring as the eyes are swelling.

Bulbs

True bulbs do not have growth eyes on their upper surface, instead they produce foliage and roots from a flattened area at the base of the bulb known as the basal plate. There are two types of true bulbs, tunicate and scaly, which are differentiated by the way their tissue layers, or scales, are arranged. Tunicate bulbs, such as onions and daffodils, have a papery outer skin (the tunica) with scales arranged in concentric rings that radiate from the centre of the bulb. Scaly bulbs, such as lilies, do not have a papery outer layer, and have smaller overlapping scales, a bit like fish scales, that are usually arranged in a spiral pattern radiating from the centre of the bulb.

Tunicate bulbs may be lifted and dried off completely, as they are protected from drying by their papery covering and they form new roots each year after dormancy. Scaly bulbs cannot withstand excessive drying because they have no protective covering and their roots, which live for more than one season, will wither if dried.

All bulbs multiply naturally to eventually form clumps, so the obvious method of propagation is to break up the bulb clumps when they are dormant. The separate bulbs are then grown on as individual plants. Small bulbs may take a couple of years to flower. Some bulbs, particularly certain lilies, form clusters of small bulblets where the stalk emerges from the bulb. These can be removed and grown on, although they may take some time to flower.

You can encourage most lilies to form bulblets or bulbils in the leaf axils along their stems by removing the flower buds and the lower leaves, then mounding up the soil around the lower part of the stem. Numerous bulblets will then form. The best of them will be those around the base, under

Bulb scooping

Method 1: cut shallow wedges out of the cut base of a bulb

the cuts will open further as the bulb shrivels and bulblets will form

Method 2: hollow out the base of the bulb

bulblets will form on the cut surface

once the bulblets have formed, place the bulb upside down in a pot, just covering the bulblets with soil

the mounded soil. A more extreme method is to disbud the stem and wrench it from the bulb, leaving the bulb in the ground. For the best results there should be a few roots at the base of the stem. Lay the stem in a shallow trench and cover it with fine soil that will not form a hard crust. You can leave a little foliage exposed, although over time a few small leaves will poke through anyway. At the end of the season the stem can be lifted and, with luck, there will be bulblets along its length.

Many bulbs can be encouraged to form bulblets by damaging their scales. This is known as scaling, and encourages bulblets to form at the point of injury, much like the cormels that form where the old corm separates from the new.

Scaling should be done at the end of the

growing season when the bulbs are plump and fleshy. Scaly bulbs are easily divided into individual scales; you can remove a few or strip the whole bulb if you wish. Tunicate bulbs must be cut up; cut down through the bulb so that each section includes a piece of the basal plate; eight sections is usually about as far as you should go. Stir the divisions up in some moist seed-raising mix in a plastic bag or wide-mouthed jar (but don't over-pack the container). Store it in a warm, moist place such as an airing cupboard. Within 8-12 weeks there will be one or more small bulblets at the basal end of each scale or division. Do not remove the bulblets, but pot up each division with the attached bulblets still in place.

Another method, bulb scooping, is really only practical for firm tunicate bulbs, such as hyacinths. Firstly, using a spoon or knife, scoop out the basal plate of the bulb. Make sure that every scale is wounded. Dust the cut surface with a fungicide and place the bulb upside down on a wire rack and store in a warm, dark place. It is important that the bulb is warm (18–21°C), but it must not dry out, so mist it occasionally. In a few days the tissue of the cut surface will contract, pulling in the outer edges of the cut. This is quite normal and not a sign of excessive drying. At this stage the temperature should be increased to around 25–30°C and in about 8-12 weeks small bulblets will form on the cut surface. The bulb can now be potted or planted out, still upside down, with the bulblets at or just below soil level.

The bulblets will come into growth in spring while the parent bulb gradually withers beneath them. Once they become dormant in autumn they may be lifted and treated as individuals.

Scoring is a simplified version of scooping. Make two intersecting wedge-shaped cuts of about 5–10 mm depth across the basal plate of the parent bulb, then proceed exactly as for scooped bulbs.

Chapter 8

Table 3: PROPAGATION BY DIVISION

Veronica teucrium

THE following table gives information about the size and type of division to use, when to divide, and the period between dividing cycles.

The size — small, medium or large — is in relation to the size of the overall clump. In other words, if large divisions are recommended, then a clump may yield only three or four new clumps, but if small divisions are acceptable then it may be possible to make 20 or 30 new clumps.

The season refers to the optimum time in a reasonably mild climate; one that experiences winter frosts, but where the soil does not freeze solid for days at a time.

The period to maturity is the time that it will take for the division to develop into a medium-sized clump, and the cycle is the time that it will take for the divisions to reach the size at which they will themselves be suitable for dividing.

Table 3 Division

Plant	Season	Size	Mature	Strike	Cycle
Acaena	any	natural layer	6-9 months	good	yearly
Acanthus	winter	large	1 year	good	biennially
Achillea	winter	small	1 year	good	yearly
Achimenes	autumn	small/medium	6-9 months	good	yearly
Aconitum	winter	medium	1-2 years	good	yearly
Acorus	any	small/medium	1 year	good	3-4 years
Adiantum	win-spr	small/medium	1-2 years	good	yearly
Adonis	sum-win	small clumps	3-6 months	good	2-3 years
Aethionema	winter	natural layer	1-2 years	good	2-4 years
Agapanthus	any	bulbous	2 years	good	2-3 years
Agastache	spring	small clump	2-3 months	good	2-3 years
Agave	any	rosette	2-3 years	good	yearly
Ajuga	aut-spr	small/medium	3-9 months	good	3-4 years
Alcea (hollyhock)	win-spr	large	1 year	moderate	1-2 years
Alchemilla	win-spr	small clump	3-6 months	good	3-4 years
Allium-ornamental	winter	bulbous	1-2 years	good	1-2 years
Aloe	any	rosette	1-2 years	good	biennially
Alstroemeria	autumn	medium	1-2 years	good	2-3 years
Alyssum	win-spr	few rosettes	3-6 months	moderate	2-3 years
Anacyclus	aut-spr	medium clump	6-9 months	good	1-2 years
Anagallis	win-spr	small clump	2-4 months	good	1-2 years
Anchusa	aut-spr	medium	1 year	good	yearly
Anemone	aut-spr	few tubers	1-2 years	good	2-3 years
Anigozanthus	aut-spr	medium	1-2 years	moderate	2-4 years
Anthemis	any	few rosettes	3-6 months	good	yearly
Anthericum	win-spr	medium	6-9 months	good	biennially
Aquilegia	win-spr	large	6 months	good	biennially
Arabis	any	natural layer	3-6 months	good	yearly
Arctotheca	spr-aut	natural layer	3-6 months	good	yearly
Arctotis	spr-aut	natural layer	3-6 months	good	1-2 years
Arenaria	any	natural layer	3-6 months	good	6 months
Arisarum (mouse plant)	aut-spr	1 good tuber	3-9 months	good	1-2 years
Aristea	spring	small clump	3-6 months	good	yearly
Armeria	any	small clump	3-6 months	good	6 months
Armoracia (horseradish)	aut-spr	1 firm root	3-6 months	good	6-9 months
Artemisia	winter	small/medium	6-9 months	good	biennially
Arthropodium	win-spr	medium clump	1 year	good	2-3 years
Arum-true forms	summer	medium tuber	1-2 years	good	biennially
Aruncus (goat's beard)	win-spr	medium clump	1-2 years	good	2-3 years
Arundinaria (bamboo)	any	small clump	3-6 months	good	yearly
Arundo	win-spr	medium clump	1-2 years	good	2-3 years
Asparagus -ornamental	win-spr	medium clump	1-2 years	moderate	2-3 years
Asphodeline	win-spr	medium	1-2 years	moderate	2-3 years

Plant	Season	Size	Mature	Strike	Cycle
Aspidistra	any	medium clump	1-2 years	good	3-4 years
Asplenium	win-spr	medium clump	1-3 years	good	3-4 years
Astelia	aut-spr	medium clump	6-12 months	good	2-3 years
Aster	win-spr	small/medium	1-2 years	good	2-3 years
Astilbe	win-spr	small/medium	1-2 years	good	1-2 years
Astrantia	aut-spr	small clump	3-6 months	good	2-3 years
Athyrium	win-spr	small clump	6-9 months	good	2-3 years
Aubrieta	win-spr	medium/large	6-9 months	moderate	biennially
Aurinia	win-spr	few rosettes	3-6 months	good	1-2 years
Bambusa (bamboo)	any	small clump	3-6 months	good	yearly
Baptisia	winter	small/medium	6-9 months	good	1-2 years
Begonia-tuberous	spring	small	3-6 months	good	1-2 years
Belamcanda	aut-win	medium	6-9 months	good	2-3 years
Bellis	win-spr	small clump	3-6 months	good	yearly
Bergenia	aut-win	medium	3-9 months	good	1-2 years
Beschorneria	aut-spr	few rosettes	6-12 months	good	2-3 years
Billbergia	spring	1-2 rosettes	6-9 months	good	2-3 years
Blechnum	win-spr	medium clump	6-9 months	good	1-2 years
Bletilla	winter	medium clump	1 year	good	1-2 years
Boltonia	winter	small/medium	3-6 months	good	1-2 years
Bomarea	winter	medium clump	6-9 months	moderate	2-3 years
Bravoa	winter	medium	1 year	moderate	2-3 years
Briza (snakegrass)	spr-aut	small clump	2-3 months	good	yearly
Bromeliads	win-spr	1-2 rosettes	6-9 months	good	2-3 years
Brunnera	aut-spr	medium	1 year	good	1-2 years
Bulbinella	win-spr	small clump	1 year	good	2-3 years
Bupthalmum	win-spr	small clump	6-9 months	good	1-2 years
Calamintha (calamint)	aut-spr	small clump	3-6 months	good	yearly
Calluna	aut-spr	natural layer	9-12 months	good	1-2 years
Caltha	aut-win	medium	1 year	good	2-3 years
Campanula	win-spr	small/medium	6-9 months	good	1-2 years
Canarina	sum/aut	1 good tuber	3-9 months	good	1-2 years
Canna	win-spr	medium	1 year	moderate	2-3 years
Cardamine	aut-spr	small clump	2-6 months	good	yearly
Cardoon	win-spr	medium	6-9 months	good	1-2 years
Carex	any	small clump	2-6 months	good	yearly
Catananche	win-spr	small/medium	6-9 months	good	1-2 years
Cattleya	postbloom	pseudobulbs	6-18 months	good	2-3 years
Celmisia	win-spr	large	1 year	Poor	3-4 years
Centauria	win-spr	small/medium	6-9 months	good	2-3 years
Cephalaria	winter	medium	1 year	good	2-3 years
Cerastium	aut-spr	small/medium	6-9 months	good	1-2 years
Chamaemelum (chamomile)	any	1-2 rosettes	3-6 months	good	yearly
Chlorophytum	spr-aut	rosette	1 year	good	yearly
Chrysanthemum	win-spr	small/medium	6-9 months	good	1-2 years
Cimicifuga	win-spr	medium clump	6-9 months	good	1-2 years
Clematis	win-spr	small/medium	6-9 months	good	1-2 years

Plant	Season	Size	Mature	Strike	Cycle
-herbaceous	win-spr	medium/large	1-2 years	moderate	3-4 years
Clivia	spring	medium clump	1 year	good	2-3 years
Colchicum	spr-sum	bulbous	1 year	good	2-3 years
Convallaria	aut-win	small clump	1 year	good	2-3 years
Coreopsis	win-spr	small clump	3-6 months	good	yearly
Coronilla					
-ground cover	spring	natural layer	3-6 months	good	yearly
Cortaderia	win-spr	small clump	1 year	good	2-3 years
Corydalis	aut-spr	small/medium	1 year	good	2-3 years
Cosmos					
-perennial forms	spring	1 tuber	1 year	good	2-3 years
Cotula	win-spr	small clump	3-6 months	good	yearly
Crinum	win-spr	bulbous	1-2 years	good	3-4 years
Crocosmia	win-spr	bulbous	1 year	good	2-3 years
Crocus	sum/aut	bulbous	1 year	good	3-4 years
Cymbidium	postbloom	pseudobulb	1-2 years	good	2-3 years
Cymbopogon (lemon grass)	spr-sum	small clump	2-3 months	good	
Cynara (globe artichoke)	spring	offset,sucker	6-9 months	good	1-2 years
Cynodon (Bermuda grass)	spr-aut	small sprig	3-6 months	good	yearly
Cyperus (papyrus)	any	small clump	3-6 months	good	1-2 years
Cypripedium	postbloom	pseudobulbs	1-2 years	moderate	3-4 years
Cyrtanthus	aut-win	bulbous	1 year	good	2-3 years
Daboecia	winter	natural layer	1 year	good	2-3 years
Dahlia	spring	1 tuber	1 year	good	1-2 years
Davallia	spring	small/medium	6-9 months	good	1-2 years
Delphinium	win-spr	medium	3-6 months	good	1-2 years
Dianella	win-spr	small/medium	1 year	good	2-3 years
Diascorea (yam)	aut-spr	1 good tuber	6-9 months	good	yearly
Dicentra	win-spr	small/medium	3-6 months	good	1-2 years
Dichondra	any	small plug	3-6 months	good	yearly
Dictamnus	winter	large	1-2 years	moderate	3-10 years
Dierama	aut-spr	small clump	1 year	good	2-3 years
Dietes syn Moraea	winter	small clump	6-9 months	good	1-2 years
Digitalis (foxglove)	aut-spr	medium/large	6-9 months	good	2-3 years
Dodecatheon	winter	medium	1-2 years	moderate	3-5 years
Doodia	win-spr	small clump	6-12 months	good	2-3 years
Doronicum	win-spr	small/medium	6-9 months	good	1-3 years
Draba	spring	few rosettes	6-9 months	moderate	2-3 years
Dracunculus	win-spr	medium	1 year	good	2-3 years
Echinacea	any	rosette	3-6 months	good	yearly
Echinops	win-spr	small clump	3-6 months	good	1-2 years
Echinopsis	any	medium/large	1 year	good	2-3 years
Epidendrum	postbloom	offsets	1-2 years	moderate	yearly
Epimedium	win-spr	pseudobulbs	1-2 years	good	1-2 years
Eranthis	autumn	small clump	6-9 months	good	yearly
Eremurus	aut-win	medium clump	3-6 months	good	2-3 years

Plant	Season	Size	Mature	Strike	Cycle
Erica	aut-spr	natural layer	9-18 months	good	1-2 years
Erigeron	win-spr	small clump	3-6 months	good	yearly
Erodium	any	small clump	2-3 months	good	yearly
Erythronium	aut-win	medium pieces	1-2 years	good	2-3 years
Eucomis	winter	small clump	1 year	good	2-3 years
Eupatorium	win-spr	medium clump	3-6 months	good	biennially
Festuca -ornamental	any	small clump	2-3 months	good	6-9 months
Filipendula	winter	medium piece	6-9 months	good	2-3 years
Fragraria	any	natural layer	3-6 months	good	yearly
Francoa	winter	small clump	3-6 months	good	1-2 years
Freesia	Summer	few rosettes	6-9 months	good	2-3 years
Fritillaria	aut-win	bulbous	1-2 years	moderate	3-5 years
Gaillardia	win-spr	bulbous	2-4 years	good	yearly
Galega	aut-spr	small clump	3-6 months	good	1-2 years
Galium (sweet woodruff)	aut-win	small clump	6-9 months	good	1-2 years
Gaultheria	win-spr	natural layer	1-2 years	good	1-2 years
Gaura	winter	small clump	6-9 months	good	1-2 years
Gazania	aut-spr	medium clump	3-6 months	good	yearly
Gentiana	aut-spr	medium clump	1-2 years	moderate	3-4 years
Geranium	any	medium clump	3-6 months	good	1-2 years
Gerbera	aut-spr	medium clump	3-6 months	moderate	2-3 years
Geum	win-spr	small clump	3-6 months	good	yearly
Glechoma	any	natural layer	2-3 months	good	6-9 months
Gloriosa	spring	1 good tuber	1 year	moderate	1-2 years
Glycyrrhiza (liquorice)	aut-spr	small clump	3-9 months	good	1-2 years
Gunnera	win-spr	small clump	2-3 years	good	3-4 years
Gypsophila	autumn	medium clump	3-9 months	good	biennially
Haemanthus		bulbous	1-2 years	good	2-3 years
Haworthia	any	few rosettes	6-9 months	good	yearly
Hedera (ivy)	any	natural layer	6-9 months	good	yearly
Helenium	any	small clump	3-6 months	good	yearly
Helianthus	win-spr	small clump	3-6 months	good	yearly
Helianthus (artichoke)	spring	1 good tuber	6-9 months	good	yearly
Helichrysum -perennial	any	medium clump	3-6 months	good	1-2 years
Heliopsis	win-spr	small clump	3-6 months	good	yearly
Helleborus	aut-spr	medium clump	1-2 years	good	3-4 years
Hemerocallis	winter	small clump	1-2 years	good	2-3 years
Hepatica	aut-win	small clump	1-2 years	good	2-3 years
Hermodactylus	aut-win	small clump	1-2 years	good	2-3 years
Herniara	win-spr	natural layer	2-3 months	good	6-9 months
Herpolirion	any	medium clump	6-9 months	moderate	2-3 years
Heterocentron syn Heeria	any	natural layer	6-9 months	good	yearly
Heuchera	aut-spr	1-2 rosettes	6-9 months	good	yearly
Hibiscus-perennial	winter	large piece	1-2 years	moderate	3-4 years

Plant	Season	Size	Mature	Strike	Cycle
Monarda	aut-spr	small clump	3-6 months	good	biennially
Montbretia	winter	bulb clump	6-9 months	good	1-2 years
Musa	spr-sum	sideshoots	6-18 months	good	1-2 years
Muscari	aut-win	small clump	6-9 months	good	biennially
Myosotis-perennial	aut-spr	small clump	3-6 months	good	1-2 years
Myrrhis (sweetcicely)	winter	small clump	3-6 months	good	1-2 years
Nelumbo	spring	firm tuber	3-9 months	good	2-3 years
Nepeta	any	small clump	2-3 months	good	6-9 months
Nephrolepis	any	medium clump	3-6 months	good	biennially
Nicotiana	spring	medium clump	3-6 months	good	1-2 years
Nierembergia	spring	small clump	3-6 months	moderate	yearly
Nymphaea	spring	medium piece	2-4 months	good	yearly
Odontoglossum	postbloom	medium clump	6-9 months	good	2-3 years
Oenothera	winter	medium clump	1-3 years	moderate	3-4 years
Omphalodes	win-spr	small clump	3-6 months	good	biennially
Ophiopogon	win-spr	small clump	3-6 months	good	1-2 years
Origanum (marjoram)	any	small clump	6-9 months	good	1-2 years
Orthrosanthus	aut-spr	small clump	2-3 months	good	yearly
Ourisia	spring	small clump	3-6 months	good	yearly
Oxalis	win-spr	medium clump	3-9 months	moderate	2-3 years
Pachysandra	aut-spr	small pieces	3-6 months	good	yearly
Paeonia -herbaceous	aut-spr	medium clump	6-12 months	good	1-3 years
Paeonia-tree	win-spr	layer, suckers	1-3 years	moderate	2-4 years
Papaver-perennial	spr-aut	medium clump	3-6 months	good	2-4 years
Paphiopedilum	aut-spr	pseudobulbs	6-18 months	good	biennially
Parahebe	postbloom	clump, layer	3-6 months	good	2-3 years
Pellaea	any	small clump	3-9 months	good	yearly
Penstemon	aut-spr	small clump	3-9 months	good	1-2 years
Perovskia	win-spr	large clump	3-6 months	good	biennially
Philodendron	winter	large clump	6-12 months	moderate	2-3 years
Phlomis	spr-sum	small clump	3-6 months	good	2-3 years
Phlox paniculata	win-spr	medium clump	3-6 months	good	1-2 years
Phormium (N.Z. flax)	win-spr	medium clump	6-12 months	good	biennially
Phygelius	aut-spr	small clump	2-3 months	good	2-3 years
Physalis	win-spr	small clump	3-6 months	good	yearly
Physostegia	winter	small clump	3-9 months	good	1-2 years
Platycodon	winter	large pieces	6-9 months	good	biennially
Pleione	win-spr	few pseudos	6-12 months	moderate	2-3 years
Polemonium	winter	small clump	3-6 months	good	2-3 years
Polianthes (tuberose)	winter	sideshoots	6-18 months	good	1-2 years
Polygonatum	aut-win	small pieces	3-9 months	good	2-3 years
Polygonum	any	natural layer	3-6 months	good	yearly
Polypodium	aut-spr	small clump	6-9 months	good	1-2 years
Polystichum	aut-spr	clump, offsets	6-9 months	good	1-3 years
Pontederia	win-spr	small clump	3-6 months	good	yearly

Plant	Season	Size	Mature	Strike	Cycle
Hosta	win-spr	small clump	3-6 months	good	1-2 years
Humulus (hop)	win-spr	medium clump	3-6 months	good	1-2 years
Hyacinthoides syn Scilla, Endymion	sum/aut	bulbous	1 year	good	2-3 years
Hypericum -ground cover	any	natural layer	6-9 months	good	yearly
Hypolepsis	any	clump, rhizome	3-6 months	good	yearly
Hyssopus (hyssop)	win-spr	small clump	3-6 months	good	yearly
Iberis-perennial	winter	large piece	6-9 months	moderate	1-2 years
Incarvillea	aut-win	medium clump	6-9 months	moderate	1-2 years
Indigofera (indigo)	spring	large clump	1-2 years	moderate	2-4 years
Inula	win-spr	medium clump	9-12 months	moderate	1-2 years
Ipheion	aut-spr	small clump	6-9 months	good	1-2 years
Ipomoea	winter	few tubers	6-9 months	good	yearly
Iris	aut-win	medium	1-2 years	good	2-3 years
Kniphofia	spring	medium clump	1-2 years	moderate	2-3 years
Lamium	any	natural layer	2-3 months	good	6-9 months
Laurentia	any	natural layer	3-6 months	good	yearly
Leonotis	win-spr	medium clump	6-9 months	good	biennially
Leontopodium	win-spr	medium clump	3-6 months	moderate	biennially
Leonurus (motherwort)	aut-spr	medium clump	3-6 months	good	1-2 years
Liatris	winter	medium clump	1 year	moderate	2-3 years
Libertia	any	small clump	1 year	good	2-3 years
Ligularia	winter	medium	1 year	good	2-3 years
Limonium syn Statice	spring	few rosettes	1 year	good	2-3 years
Linaria	winter	small clump	6-9 months	good	biennially
Linnaea	winter	medium clump	6-9 months	moderate	1-2 years
Linum-perennial	winter	small clump	6-9 months	good	1-2 years
Liriope	any	small clump	6-9 months	good	1-2 years
Littonia	winter	1 good tuber	1 year	moderate	biennially
Lobelia-perennial	aut-spr	offsets	1-2 years	good	2-3 years
Lobivia	any	offsets	1-2 years	good	yearly
Lotus	spr-aut	natural layer	3-6 months	good	yearly
Lupinus (lupin)	win-spr	small clump	6-9 months	good	biennially
Lychnis	win-spr	small clump	1 year	good	biennially
Lysichiton	any	large piece	1-2 years	poor	3-5 years
Lysimachia	any	natural layer	3-6 months	good	biennially
Lythrum	winter	large clump	6-9 months	moderate	biennially
Macleaya	winter	small clump	6-9 months	good	1-2 years
Mammillaria	any	offsets	1-2 years	good	yearly
Marrubium	aut-spr	small clump	2-3 months	good	yearly
Mazus	any	natural layer	3-6 months	good	yearly
Melissa (balm)	aut-spr	small clump	3-6 months	good	6-9 months
Mentha (mint)	any	small clump	2-3 months	good	6 monthly
Mertensia	aut-win	small clump	1 year	good	2-3 years
Mimulus -ground cover	any	small clump	2-3 months	good	6-9 months
Mirabilis	win-spr	1 good tuber	6-9 months	good	1-2 years

Plant	Season	Size	Mature	Strike	Cycle
Potentilla	any	runners, layer	6-12 months	good	1-2 years
Poterium (burnet)	aut-spr	few rosettes	3-4 months	good	6-9 months
Pratia	any	clump, layer	2-3 months	good	6-9 months
Primula	postbloom	small clump	3-9 months	good	1-2 years
Primula (polyanthus)	sum/aut	rosette	3-6 months	good	1-2 years
Pteris	aut-spr	small clump	3-9 months	good	1-2 years
Pterostylis	postbloom	pseudobulbs	9-18 months	moderate	2-3 years
Pulmonaria	aut-spr	small clump	3-6 months	good	yearly
Pulsatilla	aut-spr	small clump	2-6 months	good	1-2 years
Ranunculus	aut-spr	small clump	3-9 months	good	1-2 years
Raoulia	spring	medium pieces	2-3 months	good	yearly
Rehmannia	win-spr	small clump	3-6 months	good	1-2 years
Reinwardtia	spring	medium pieces	3-6 months	good	1-2 years
Rhododendron -evergr. azalea)	aut-spr	natural layer	9-12 months	good	1-2 years
Rhododendron -low	aut-spr	natural layer	1-2 years	moderate	2-3 years
Rhodohypoxis	win-spr	small clump	3-6 months	good	2-3 years
Rhoeo	aut-spr	rosette	3-6 months	good	1-2 years
Rhubarb	win-spr	medium clump	1-2 years	good	yearly
Rhus	aut-spr	suckers	1-2 years	good	1-2 years
Rodgersia	win-spr	small clump	6-12 months	good	2-3 years
Romulea	aut-spr	bulbous	1-2 years	good	3-4 years
Rubia (madder)	win-spr	medium clump	6-9 months	moderate	biennially
Rubus-ornamental	any	natural layer	3-6 months	good	yearly
Rudbeckia -perennial	win-spr	small clump	3-6 months	good	yearly
Rumex (sorrel)	win-spr	medium clump	2-3 months	good	yearly
Ruscus	win-spr	suckers	6-18 months	good	2-3 years
Ruta (rue)	win-spr	medium clump	2-4 months	good	yearly
Sagina	any	small clump	2-3 months	good	6-9 months
Salvia-perennial	aut-win	small clump	3-6 months	good	1-2 years
Sandersonia	win-spr	1 good tuber	6-9 months	good	1-2 years
Sanguinaria	aut-win	small pieces	3-6 months	moderate	1-3 years
Sanguisorba syn Poterium	winter	medium clump	3-6 months	moderate	1-3 years
Sanseveria	spr-aut	rosette	6-9 months	good	2-3 years
Saponaria (soapwort)	any	natural layer	2-3 months	good	6-9 months
Sasa (bamboo)	any	small clump	3-6 months	good	yearly
Satureja (savoury)	aut-spr	large clump	2-3 months	good	1-2 years
Saxifraga	any	few rosettes	2-6 months	moderate	2-3 years
Scabiosa	win-spr	large clumps	3-6 months	good	1-3 years
Schizostylis	aut-spr	small clump	3-6 months	good	1-2 years
Scilla	sum/aut	bulbous	3-6 months	good	1-2 years
Scirpus	any	small clump	2-3 months	good	6-9 months
Scutellaria	win-spr	large clump	6-12 months	moderate	2-4 years
Sedum	any	natural layer	2-6 months	good	6-9 months
Sempervivum	any	rosette	2-6 months	good	yearly

Plant	Season	Size	Mature	Strike	Cycle
Senecio -not shrubby	winter	medium clump	2-6 months	good	1-2 years
Shortia	winter	runner, sucker	6-9 months	good	2-3 years
Sidalcea	winter	small clump	3-6 months	good	1-2 years
Silene	aut-spr	medium clump	3-9 months	moderate	2-3 years
Sisyrinchium	any	small clump	3-6 months	good	yearly
Smilacina	winter	small pieces	3-9 months	good	2-3 years
Solanum-tuberous	win-spr	1 good tuber	6-9 months	good	yearly
Solidago (goldenrod)	win-spr	small clumps	3-9 months	good	1-3 years
Soleirolia	any	small clump	1-3 months	good	3-9 months
Stachys	aut-spr	medium clump	3-6 months	good	2-3 years
Sternbergia	sum/aut	bulbous	6-9 months	good	2-3 years
Stokesia	winter	small clump	3-6 months	good	1-2 years
Strelitzia	aut-spr	medium clump	1 year	good	3-4 years
Streptocarpus	aut-spr	few rosettes	3-9 months	moderate	biennially
Symphytum (comfrey)	aut-spr	medium clump	2-4 months	good	1-2 years
Syringa -non-grafted	aut-spr	suckers	2-6 months	good	1-2 years
Tanacetum (tansy)	aut-spr	small clump	2-6 months	good	yearly
Tellima	aut-spr	small clump	3-9 months	good	1-2 years
Tetrapanax	spring	offset, sucker	6-18 months	good	2-3 years
Thalictrum	aut-win	medium clump	6-9 months	good	2-3 years
Thymus (thyme)	any	small clump	2-6 months	good	yearly
Tolmiea	any	natural layer	3-9 months	good	yearly
Tradescantia	any	small clump	3-9 months	good	1-2 years
Tragopogon (salsify)	win-spr	small clump	3-6 months	good	yearly
Tricyrtis	win-spr	small clump	6-9 months	good	biennially
Trillium	aut-win	medium pieces	3-9 months	moderate	2-4 years
Trollius	win-spr	small clump	3-6 months	good	1-3 years
Tropaeolum	win-spr	few tubers	6-9 months	good	1-3 years
Tulbaghia	summer	medium clump	3-6 months	good	1-2 years
Urceolina	aut-win	bulbous	3-6 months	good	2-3 years
Valeriana	aut-spr	small clump	3-6 months	good	1-2 years
Vancouveria	aut-win	medium clump	3-6 months	good	2-3 years
Verbena	aut-win	clump, layer	2-3 months	good	6-9 months
Veronica	any	small clump	3-6 months	good	1-2 years
Viola	aut-spr	clump, layer	2-3 months	good	6-9 months
Viscaria-perennial	aut-spr	small clump	3-6 months	good	1-2 years
Wachendorfia	winter	few tubers	6-9 months	good	1-3 years
Wahlenbergia	aut-spr	firm rhizome	3-9 months	good	biennially
Yucca	spring	few rosettes	6-9 months	moderate	2-3 years
Zantedeschia (calla)	win-spr	firm rhizome	3-6 months	good	2-3 years
Zauschneria	winter	small clump	3-9 months	good	2-3 years
Zingiber (ginger)	win-spr	firm rhizome	3-6 months	good	1-2 years
Zygopetalum	postbloom	medium clump	6-12 months	good	2-3 years

Chapter 9

PROPAGATION BY CUTTINGS

Fuchsia 'Arcady'

TAKING a cutting involves removing a piece of tissue, stem, leaf or root from a parent plant and inducing it to form roots or foliage of its own so that it can be grown on as a new plant. You need to grow plants from cuttings in order to propagate most of the huge range of woody-stemmed plants that is now available to gardeners. If you don't have this ability, your propagation of trees and shrubs will be limited to species and those few varieties that come relatively true to type from seed.

Most woody-stemmed plants may be propagated from cuttings, but some do not grow well on their own roots and may have to be budded or grafted onto vigorous rootstocks.

In many cases the stock for the graft is grown from a cutting, so a knowledge of propagation from cuttings is important for grafting and budding too (see chapter 11 for budding and grafting techniques).

Cutting stock plants

Most beginners use their garden plants as cutting stock. However, you'll often get better results by keeping special stock plants or by preparing your garden plants in advance for use as cutting stock.

The best cuttings come from vigorous new growth; pruning back an intended cutting parent a couple of months before taking cuttings will encourage strong new shoots to develop. Of course you can always use your prunings as cuttings too.

If you find yourself regularly propagating the same plants, it is a good idea to pot up a few and keep them as stock plants. That way they can be fed and pruned to produce the maximum growth and if you want cuttings out of season they can be kept growing indoors.

Wounding

Cuttings usually produce roots from the cut end at the base of the stem. However, some plants do better if more cambium is exposed. Making a wound on the side of the stem, either as a thin cut or by removing a sliver of bark, may encourage these hard-to-strike plants to develop roots. Some plants, especially rhododendrons, are routinely wounded when taking cuttings, but for most it is something to try if they seem reluctant to strike.

Root-forming hormones

Plant growth and reproduction is regulated by agents known as phytohormones. They are not true hormones, as found in animals, but they act in similar ways.

There are four groups of phytohormones: auxins, gibberellins, cytokinins and inhibitors. The best known are the auxins, such as indolebutyric acid, which is found in the compounds often sold as root-forming hormones. Auxins can promote growth or inhibit it. In stems, auxins promote cell elongation and cell differentiation, whereas in roots they inhibit growth in the main roots but promote the development of adventitious roots. Auxins form at the tips of stems or roots then flow back through the rest of the plant.

Most cuttings are capable of producing enough of these auxins to stimulate root development, but it sometimes takes quite some time before the roots form. We can speed up this process and get better root development by applying the auxin directly

Dipping a softwood cutting in root-forming hormone powder will speed up the process of root development.

at the point where we want roots to form.

Root-forming hormones are available in powder, gel or liquid forms. The powder comes in varying strengths based on the intended use (softwood, semi-ripe or hardwood), while the gel and the liquid can be diluted to any required strength. Knowing what strength to use is largely a matter of experience, as several variables, such as the type of plant, state of growth, time of year and the propagating environment, all play a part. In most cases root-forming hormones are not essential, but they can make a difference and will tilt the odds of success a little more in favour of the propagator.

Stem cuttings

The main types of stem cuttings are herbaceous (or greenwood), softwood, semi-ripe

New growth suitable for softwood cuttings.

(or semi-hardwood) and hardwood. By selecting the appropriate method it is possible to have cuttings in production year round.

Most stem cuttings include the growing tip of the branch from which they are taken, which can be a main growing tip or a side shoot. Stem cuttings need not include a growing tip if the wood is of the right type, but the closer the apical meristem (the natural auxin source), the faster the cutting will strike.

Softwood and semi-ripe cuttings

Most perennials and many shrubs and trees, both evergreen and deciduous, can be grown by using cuttings of soft new tip growth. Such softwood cuttings have the advantage of being taken from the most actively growing part of the plant. This means that they strike quickly and develop good root systems, but there are a few disad-

vantages too. The very soft shoots are quite tender and can be damaged in the act of taking the cutting and, unless you have a misting system, they tend to wilt and rot before they strike.

Softwood cuttings are very popular with commercial growers, who usually have sophisticated equipment; home gardeners may have more success with semi-ripe cuttings. There is no clear-cut dividing line between these two types of cutting; semi-ripe cuttings are just softwood cuttings that have matured a little. In most cases the only difference between the two types of cutting is the time of year at which they are taken. Softwood cuttings tend to be most readily available in spring and early summer, while semi-ripe cuttings are more commonly taken in late summer and autumn.

The method of preparing the cuttings is fairly standard and does not vary much from plant to plant. The most important factors are timing and the size of the cutting. Precise timing is crucial in only a few cases, although every plant has an optimum time. Evergreen azaleas, for example, can be struck throughout the year, but the most successful method is to use fairly soft small cuttings under mist, and they strike best if taken in the short period between mid spring to early summer.

The size of the cutting is usually more important. Beginners often take cuttings that are too large and too mature. The belief seems to be that the bigger and firmer the cutting, the more likely it is to survive and hence eventually to strike. Very few cuttings need to be more than three or four nodes long. Softwood cuttings are seldom more than 75 mm long, while semi-ripe cuttings tend to be larger, typically 100–150 mm long. This is because softwood cuttings are usually taken from immature growth, whereas semi-ripe cuttings are more mature and as the growth matures so the internodal length increases. Softwood cuttings are, by their

Cuttings at various stages of preparation: (from left) untrimmed, excess foliage removed, foliage cut back and cutting ready to insert.

Fresh softwood cuttings prepared and planted.

very nature, tip cuttings, but semi-ripe cuttings may also be taken from further down the stem. Nevertheless, tip cuttings strike best, regardless of whether they are soft or semi-ripe.

Having taken the cutting off the plant, carefully strip the leaves from the two lower nodes, or just the bottom node if the cutting is very small. The softer the cutting, the more likely it is to be damaged as the leaves are removed; it is easy to strip off the bark if you just pull the leaves downwards to remove them. Most leaves come away cleanly if they are removed with an upward pulling action after being pulled downwards just enough to break the join between leaf and stem.

Once the lower leaves have been removed, trim back the remaining leaves by about half their length to reduce the transpiration area. Next insert the cuttings in a tray of fresh mix, I prefer a finely sieved 50/50 mixture of bark-based potting mix and perlite. Very soft cuttings may need to be dibbled into place to stop them being bruised, but most cuttings will not be damaged if they are gently pushed into place. Space the cuttings so that they do not overlap; this is important if they are going under mist because leaves that are hidden under other foliage will not receive any mist.

If you do not have a misting unit, cover the cutting tray with a polythene bag or keep it in a humid enclosed propagator. Under such conditions, your cuttings will need to be slightly firmer than those struck with mist because they will be more likely to suffer from moisture stress, which may cause wilting. Nevertheless, do not use heavier material than is absolutely necessary.

Bottom heat in the form of heating cables or a heated propagating tray can speed up the formation of roots, especially for cuttings taken during the colder parts of the year. Always make sure that the bottom heat is not causing excess drying of the potting mix, as softwood cuttings usually die if they are allowed to wilt.

Don't use vegetable-oil based wilt-proofing agents on your cuttings. These work by coating the leaf to stop moisture loss but they also affect the leaf's ability to transpire, which may hasten rotting. Water, perhaps with a little fungicide added, is all that is required.

Micro-cuttings

Large-leafed plants are simple to work with and present no special problems when taking cuttings. Not so those plants with tiny little needle-like leaves, like ericas and callunas, or those with very short lengths of new growth and almost no perceptible internode, such as many of the small alpine shrubs and perennials.

However, most of these plants are quite easy to propagate. You just have to use very small cuttings. I call them micro-cuttings and find them some of the most satisfying to work with because it really is a case of creating something from almost nothing. For example, a typical *Erica* or *Calluna* cutting is no more than about 15 mm long, so it is very satisfying to look at the shrub that develops and think back to what it was grown from.

Micro-cuttings are rather fiddly to work with: they tend to break very easily and can be difficult to position in the propagating trays. Use a much more finely sieved mix than you would for larger cuttings and moisten it slightly as you work. The finer, more tightly packed soil grains will hold the cuttings better and the moisture will help to bind the soil grains together.

These plants only rarely strike from anything other than pieces of tip growth. Also, flowering shoots do not, as a rule, strike well. Use the very tip pieces of the non-flowering shoots, which are usually easy to distinguish as they are more wiry and often have more widely spaced leaves. The succulent, tightly foliaged, bright green flowering shoots often look as if they would be the pieces more likely to strike, but they are not.

Greenwood and basal cuttings

Some shrubby plants, such as *Pelargonium*, do not develop true woody stems, except at the base of very old plants. These sub-shrubs are propagated by a form of softwood cutting known as a greenwood cutting. Personally, I see no difference, and make no distinction, between greenwood, softwood and semi-ripe cuttings, but as you will perhaps see the term used occasionally it may help to have it pointed out.

In spring many herbaceous and tuberous perennials, such as *Begonia*, *Dicentra* and *Delphinium*, produce vigorous, fleshy shoots that emerge directly from their root clumps.

These strong-growing basal shoots can be used as softwood cuttings. Do not allow the cuttings to get too large before using them, as they strike much faster if taken before the leaves are fully expanded. Most plants that are capable of being grown from basal cuttings develop very quickly once they burst into growth, so there is only a limited time in which to work. Taking basal cuttings is a good way to quickly build up stock without having to break up the root clump.

Heel cuttings

Heel cuttings are a form of semi-hardwood cutting. Instead of being cut from the parent stock, the cutting is removed by pulling it off so that it comes away with a flap or 'heel' of bark at the base of the cutting. The idea behind this method, which is often used with conifer cuttings, is that it firmly anchors the cutting and provides a greater cut surface on which roots may form. I find that it really has little practical value and tends to disfigure the parent stock far more than conventional cutting methods.

Leaf-bud cuttings

Climbers often have an extremely long internodal length, so traditional stem cuttings tend to be large and unwieldy. It is possible to make smaller cuttings and get more of them by using a modified form of stem cutting known as a leaf-bud cutting. Select a fairly mature leaf and cut the stem just above and below so that you have the leaf and a piece of stem about 50 mm long. Trim the leaf back to about half size, dip the base of the stem in root-forming hormone, and insert the cutting.

If the plant has double leaf buds directly opposite one another, as with *Clematis*, the stem can be split down the middle and each half, complete with leaf and bud, can be used as a cutting. These techniques can be used only with reasonably firm wood. They work best with evergreen climbers or with early season deciduous material.

Conifer cutting with heel.

Honeysuckle leaf bud cuttings.

Hardwood cuttings

Hardwood cuttings, which are taken in the winter when the plants are dormant, are generally used to propagate deciduous shrubs and trees. Many of these plants can be grown from soft or semi-hardwood cuttings taken during the growing season, but hardwood cuttings are often preferred as they can be taken without complex propagating equipment. This makes them particularly useful if you have no greenhouse or propagating unit.

Hardwood cuttings tend to be longer than semi-ripe cuttings, from 150–300 mm, but are otherwise very similar, except they have no leaves. They are usually struck outdoors in specially prepared garden beds that have been well cultivated or filled with potting mix. There is, however, no reason why a cold frame or other propagating unit cannot be used if one is available.

Many hardwood cuttings can also be struck indoors with bottom heat. Take the cuttings between mid and late winter and place them on a heated bed or in a frost-free greenhouse. This will induce early growth and they should develop roots as they sprout. While this requires more environmental control, it is quicker than striking the cuttings outdoors.

Mallet cuttings are a form of hardwood cutting that uses a small portion of the main stem and a side shoot. They are most often used when the side shoots are long and wiry and therefore likely to move around the cutting mix. The small portion of the main stem acts as an anchor that holds the wiry side-

The long internodes on these honeysuckle stems make them well suited to leaf-bud cuttings.

shoot in the cutting mix, lessens the risk of the cutting drying out and exposes more cambium for root formation. Using a section of the main shoot either side of the side shoot creates an anchor for the cutting as well as increasing the root-forming area. This method can also be used for semi-hardwood cuttings of wiry-stemmed plants, particularly climbers.

Plants with pithy stems, such as hydrangeas, are suitable for use as hardwood cuttings, but the fibrous pith may dry up or rot before the cutting strikes. To get round this problem cut immediately below a node where the stem is at its hardest and seal the bottom of the cutting with pruning paste or wax. Wound the sides of the stem to expose some of the cambium, then insert it in the cutting mix.

Conifers can be struck from hardwood cuttings, although they may take several months to form roots. Hardwood conifer cuttings will have foliage, so while it is possible to strike them in the open ground, they tend to dry out and die when exposed to sun and wind. You'll get better results using a cold frame or by enclosing the propagating area in a plastic tent, preferably in the shade.

Leaf cuttings

Leaf cuttings are an alternative that may work with plants that are not easily divided or that do not produce stems suitable for use as cuttings. They seem to be most successful with plants from tropical or very humid areas.

There are several different types of leaf cutting, the most straightforward of which is the petiole or leaf stem cutting. This is commonly used with African violets (*Saintpaulia*) and other Gesneriads. Simply remove a leaf from the plant, trim the petiole back to about 25–50 mm long and make a small hole in the potting mix for the base of the leaf. Firm the leaf into place so that it will remain standing up.

When the tray or pot is complete, cover it with a tent of clear plastic film to maintain

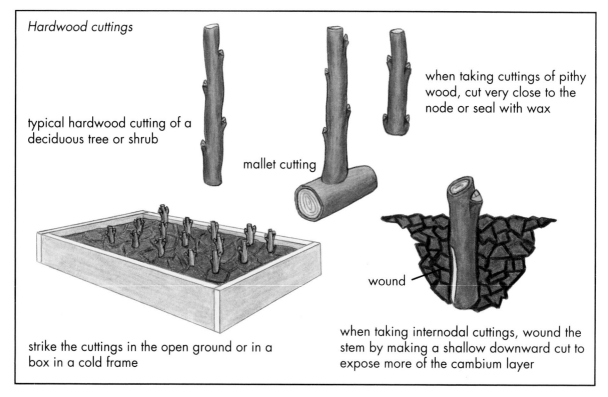

Hardwood cuttings

typical hardwood cutting of a deciduous tree or shrub

mallet cutting

when taking cuttings of pithy wood, cut very close to the node or seal with wax

wound

strike the cuttings in the open ground or in a box in a cold frame

when taking internodal cuttings, wound the stem by making a shallow downward cut to expose more of the cambium layer

the humidity and store it in a bright warm place out of direct sunlight. Bottom heat will be a definite advantage. Leaf cuttings do well in small heated propagating units.

With some plants, particularly begonias, Cape primrose (*Streptocarpus*) and *Sansevieria*, it is possible to cut up the leaf and use the sections as cuttings. This is easiest with large leaves, particularly long narrow leaves like those of *Sansevieria*. The most common method is to cut straight across the leaf and insert the cut end into the soil mix, tip end up. It is also possible to cut the midrib from the centre of the leaf and use the two halves as cuttings by inserting the cut side into the mix.

The large-leafed begonias are usually propagated by a slightly different method known as leaf slashing. Remove a healthy leaf and trim the petiole back to about 10 mm long. Next make cuts through several of the prominent veins on the underside of the leaf, making sure that you go right through the leaf. Using small pieces of wire, pin the leaf, topside up, to a tray of moist cutting mix, and cover the tray with a close-fitting polythene cover. After a few weeks, small plantlets will develop on the leaf at the cut points.

Another method is to cut the leaf into small portions and pin those to the mix. Provided each has a section of vein, they will be able to strike. Small plantlets will appear where the leaf contacts the soil. This takes 6–8 weeks depending upon the conditions. Many succulent plants can be grown by leaf cuttings too, but they are better kept uncovered in less humid conditions. However, it is important to keep the soil moist so that the points of contact between the soil and leaves do not dry out.

Some plants, such as the piggy-back plant (*Tolmiea menziesii*) and the hen and chicken fern (*Asplenium bulbiferum*), develop small plantlets along the edges of the leaves or at the point where the petiole joins the leaf.

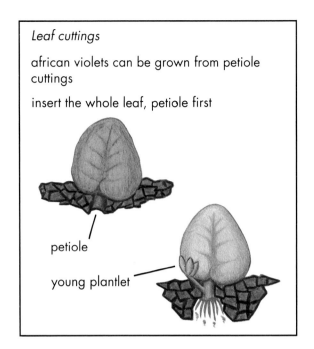

Leaf cuttings

african violets can be grown from petiole cuttings

insert the whole leaf, petiole first

petiole

young plantlet

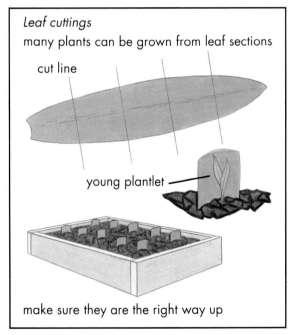

Leaf cuttings

many plants can be grown from leaf sections

cut line

young plantlet

make sure they are the right way up

Any leaves on which embryo plants develop can be removed and pinned to moist potting mix. The young plantlets will quickly develop roots where they touch the soil or as they drop from the parent leaf. Once again, some succulents, such as *Kalanchoe* (*Bryophyllum*), have similar tendencies but prefer a somewhat drier propagating environment.

Aftercare

Determining when their cuttings have struck and when they should be pricked out often perplexes beginners. It is usually easy enough to determine when a cutting has struck because it starts to make new growth, but deciding when it should be pricked out is a different matter. You don't want to do it too early or you may disturb the cuttings while they still have poorly developed root systems, but leave it too late and your cuttings will be weakened by being left in the propagating tray too long.

The ideal time to prick out a cutting is when its root mass roughly equates to the amount of foliage it is carrying. This will ensure that the plant can be self supporting. However, most plants are quite adaptable and provided they are given an occasional liquid feeding they can be left in the propagating tray for a considerable time.

Except with very small cuttings, which benefit from a finer mix, ordinary potting mix is a perfectly suitable medium for freshly struck cuttings. Don't overpot your cuttings, using pots that are too large leaves room for weeds to develop and will take up more of your valuable space. However, do consider the potential rate of growth of the cutting. Vigorous perennials would be quite happy pricked out directly into 125 mm pots, whereas an alpine rhododendron would be better in a 55-mm propagating tube.

Cuttings struck in covered frames or under mist must be gradually hardened before

A *Kalanchoe* (formerly *Bryophyllum*) leaf with young plantlets attached.

going outside or they may suffer from moisture stress when exposed to sun or wind. A week or so in a shadehouse or cool-shaded part of the garden is usually enough to toughen them up. Once the cuttings have been hardened off they can be treated as adult plants and fed and potted on as their growth rate demands.

Autumn-struck cuttings pose the problem of deciding what to do with the young plants over winter. In mild, near frost-free climates this is no great difficulty because they can still be kept outside, but the colder the climate the more likely it is that they will have to be kept indoors. If the weather is not too extreme, you may be able to gradually harden them off in a cold frame then move them outside.

Chapter 10

Table 4: PROPAGATION BY CUTTINGS

Japanese maples (*Acer*)

THE following table details the most common methods of taking cuttings for particular genera. The time it takes for a cutting to strike can vary markedly depending on the conditions. Cuttings under mist or fog will tend to take considerably less time to strike than those under polythene or in frames. Also, misted cuttings may be taken in a softer condition and over a longer period. The strike percentage is heavily influenced by the method use; generally the more primitive the conditions, the lower the strike rate.

Table 4 Propagation by cuttings

Plant	Stem Cuttings — Type of Cutting Soft	Semi	Hard	Leaf	Root	Season	Time (days)	Strike
Abelia		●				spr/sum	30	90%
Abeliophyllum		●				sum/aut	28-60	60-80%
Abies		●				sum/aut	50-80	30-75%
Abutilon	●	●				sum/aut	25-50	60-90%
Acacia		●			●	spr-aut	25-80	20-75%
Acanthus					●	winter	40-80	90%
Ackama		●				sum/aut	30-70	75%
Acmena		●				sum/aut	50	60%
Actinidia	●	●				sum/aut	20-40	60-95%
Adenandra		●				spr-aut	30-50	50-80%
Aethionema	●					spr-aut	14-35	90%
Agapetes		●				spr-aut	30-60	70-90%
Agonis		●				summer	40-60	30-60%
Ailanthus					●	spr-aut	30-45	65%
Ajuga	●					any	10-20	100%
Akebia		●				spr/sum	20-40	80%
Alberta		●				sum/aut	30-70	50-75%
Albizia		●		●		sum/aut	20-40	75%
Allamanda		●				spr-aut	25-60	60-80%
Aloe								
-climbing or bush		●				spr-aut	15-40	90%
Aloysia syn Lippia	●					sum-win	30-100	75%
Alseuosmia		●				spr-aut	30-70	80%
Alyssum								
-perennial forms	●					sum/aut	15-40	90%
Ampelopsis		●				sum-win	30-100	75%
Anchusa					●	sum-win	15-50	80%
Andromeda		●				spr-aut	30-50	60-90%
Androsace	●					spr-aut	20-30	90%
Anthemis	●					spr-aut	15-30	90%
Antigonon		●				spr-aut	30-50	60-80%
Arbutus		●				spr-aut	40-80	20-50%
Arctostaphylos		●				sum/aut	30-60	50-80%
Arctotis	●					spr-aut	20-40	70-90%
Ardisia		●				spr-aut	20-40	50-75%
Aristolochia	●					summer	20-40	70-90%
Aristotelia		●				spr-aut	25-70	60-80%
Artemisia	●	●				spr-aut	15-40	80-90%
Asclepias	●					sum/aut	20-40	40-80%
Astartea		●				spr-aut	28-60	60-80%
Aster	●					summer	10-25	90%
Aubrieta	●					any	15-40	90%
Aucuba		●				sum/aut	30-60	80%
Avocado		●				sum/aut	30-60	40-75%
Azara		●				sum/aut	40-60	30-75%
Backhousia		●				autumn	30-70	50-75%

Plant	Stem Cuttings — Type of Cutting Soft	Semi	Hard	Leaf	Root	Season	Time (days)	Strike
Baeckia		●				spr/aut	21-50	70-90%
Banksia		●				sum/aut	30-80	20-80%
Baptisia	●					spring	15-30	70-90%
Bauera	●	●				spr-aut	28-60	60-80%
Bauhinia		●				spr-aut	20-40	75%
Beaufortia		●				spr-aut	25-50	75-90%
Beaumontia		●				sum/aut	25-50	50-75%
Begonia-tuberous				●		spr-sum	15-40	80%
Bellis	●					spr-aut	10-20	95%
Berberidopsis		●				spr-aut	25-60	60-80%
Berberis		●				sum-aut	40-100	60%
Berzelia		●				spr-aut	21-42	90%
Bignonia		●				spr-aut	25-50	60-80%
Billardiera		●				spr-aut	25-70	30-75%
Boltonia	●					spr-sum	15-40	70-90%
Boronia		●				spr-aut	30-75	50-80%
Bougainvillea		●				sum/aut	30-60	30-75%
Bouvardia					win	spr-aut	25-40	75%
Brachycome	●					spr-aut	10-25	90%
Brachyglottis		●				sum-win	25-80	80%
Brachysema		●				sum/aut	20-40	30-60%
Brugmansia syn datura)	●	●				sum/aut	20-40	50-80%
Brunfelsia		●				spr-aut	20-50	75%
Buddleia	sum	aut				spr-aut	35-100	80%
Bupthalmum	●					spr-aut	10-28	90%
Buxus		●				sum/aut	35-70	40-80%
Caesalpinia		●				summer	30-50	40-75%
Calamintha	●					spr-aut	7-21	100%
Calceolaria								
-perennial forms	●					spr-aut	10-30	90%
Calliandra	sum		win			sum/aut	25-50	60-80%
Callicarpa		aut				sum/aut	20-100	40-80%
Callistemon		●				sum/aut	25-40	90%
Calluna		●				autumn	25-45	90%
Calocedrus syn Libocedrus								
Calothamnus		●				sum/aut	30-80	50-75%
Calytrix		●				sum/aut	30-60	30-60%
Camellia		●				summer	25-40	80%
Campanula	●					spr-aut	40-80	50-80%
Campsis	sum		win			summer	15-30	90%
Cantua		●				sum/aut	30-100	80-50%
Cardamine				●		spr-aut	20-50	75%
Carica (papaya, babaco)						spr-aut	10-21	100%
Carissa	●					spr-aut	15-40	90%
Carissa		●				sum/aut	25-60	20-60%
Carmichaelia		●				spr-aut	30-70	75%

Table 4: Propagation by cuttings

Plant	Stem Cuttings — Type of Cutting					Season	Time (days)	Strike
	Soft	Semi	Hard	Leaf	Root			
Carpentaria	•					sum/aut	40-100	10-50%
Caryopteris		aut	win			sum/aut	40-100	60%
Cassia	•	•				sum/aut	25-60	20-60%
Cassinia	•	•				sum/aut	25-50	40-80%
Casuarina	•	•				sum/aut	30-80	50-75%
Catananche	•			•		winter	40-70	80%
Cavendishia		•		•		sum/aut	25-50	40-80%
Ceanothus		•		•		winter	25-50	75%
Cedrela		•				winter	50-100	75%
Cedronella (balm of Gilead)	•					spr-aut	10-28	90%
Celastrus			win			win-spr	30-80	75%
Celtis		•				sum/aut	40-60	30-60%
Centaurie –perennial forms	•					sum/aut	10-30	90%
Ceratopetalum		•				sum/aut	30-60	75%
Ceratostigma	•	•				sum/aut	25-50	40-75%
Cercidiphyllum	•	•				sum/aut	30-60	40-75%
Ceropegia		•				summer	25-40	75%
Cestrum		•				autumn	25-40	75%
Chaenomeles	sum		win			sum/aut	25-60	90%
Chamaecyparis		•				sum/aut	35-70	40-75%
Chamaelaucium		•				sum/aut	25-50	30-75%
Chamaemelum (chamomile)	•					spr-aut	10-20	100%
Cheiranthus (wallflower)		•				spr-aut	15-30	80%
Choisya	•	•				spr-aut	25-50	50-85%
Chorizema	•	•				spr-aut	30-50	50-80%
Chrysanthemum	•	•				spr-aut	10-20	100%
Chrysocoma	•					any	15-30	100%
Cinnamomum		•				sum/aut	25-50	60%
Cissus	•	•				spr-aut	25-40	80%
Cistus	•	•				spr-aut	25-40	75%
Citrus		•				sum/aut	25-50	75%
Clematis	•	•				spr-aut	25-75	25-80%
Clerodendrum	•	•			win	spr-aut	30-50	20-60%
Clethra	•	•				sum/aut	25-40	50-75%
Clianthus		•				spr-aut	20-40	60-85%
Clytostoma		•				spr-aut	25-50	50-80%
Cobaea	•					spr-aut	10-25	80%
Coleonema		•				spr-aut	25-50	50-85%
Coleus	•	•				spr-aut	5-15	100%
Congea		•				spr-aut	25-60	50-75%
Convolvulus	•	•				spr-aut	14-30	85%
Coprosma		•				any	25-50	80%
Corallospartium		•				spr-aut	30-70	75%
Cordyline		•			•	any	21-60+	60-90%

Plant	Stem Cuttings — Type of Cutting					Season	Time (days)	Strike
	Soft	Semi	Hard	Leaf	Root			
Coreopsis	•					spr-aut	10-20	95%
Cornus	some	some	most			varies	varies	40-60%
Corokia		•				spr-aut	20-50	40-80%
Coronilla		•				spr-aut	14-30	80%
Correa		•				spr-aut	20-40	75%
Corylopsis			•			spr/sum	20-40	75%
Corylus			•			winter	75-100	75%
Corynocarpus		•				spr-aut	28-50	60-80%
Cosmos	•					summer	10-25	90%
Cotinus	sum	sum	win			spr-aut	30-100	30-75%
Cotoneaster	sum	sum	win				30-100	50-80%
Crassula	sum	sum	win			spr-aut	15-30	100%
Crataegus		•				spr-aut	40-100	20-50%
Crinodendron		•				spr-aut	30-60	50-75%
Crotalaria		•				spr-aut	15-40	75-90%
Crowea		•				spr-aut	25-50	60-80%
Cryptomeria		•				spr-aut	30-75	75%
Cuphea	•	•				any	10-20	100%
Cupressocyparis		•				any	40-100	75%
Cupressus		•				sum/aut	30-75	40-75%
Cyathodes		•				spr-aut	30-70	30-60%
Cyphomandra	•					spr-aut	10-30	80%
Cytisus	sum	sum	win			spr-aut	30-80	80%
Daboecia		•				spr-aut	20-40	85%
Dacrydium		•				autumn	40-80	10-60%
Dahlia	•					spring	10-28	90%
Dampiera		•				spr-aut	20-50	70-90%
Daphne		•				spr-aut	30-75	40-80%
Davidia		•				sum-aut	30-70	30-65%
Delphinium	•					spring	10-25	90%
Desfontainea		•				spr-aut	25-50	40-75%
Deutzia	sum	sum	win			spr-aut	30-100	75%
Dianthus		•				spr-aut	15-30	60-90%
Diascia	•					spr-aut		
–perennial forms	•					spr-aut	10-30	100%
Dicentra						spr-sum	10-25	90%
Digitalis						spring	10-25	90%
Dimorphotheca						spring	15-30	75-90%
Diosma		•				spr-aut	25-50	60-85%
Distictis syn Phaedranthus						spr-aut	15-40	85%
Dizygotheca		•				spr-aut	20-50	80%
Dodonea		•				spr-aut	30-60	75%
Dolichos		•				spr-aut	15-40	90%
Dombeya		•				spr-aut	15-30	90%
Dracaena		•				spr-aut	30-60	85%
Dracophyllum		•				spr-aut	40-100	20-60%
Dregia	sum		win			spr-aut	30-80	50-75%
Drimys		•				spr-aut	20-50	50-75%

79

Type of Cutting

Stem Cuttings Plant	Soft	Semi	Hard	Leaf	Root	Season	Time (days)	Strike
Drosanthemum	•	•				spr-aut	14-28	70-90%
Dryandra	•	•				spr-aut	30-70	30-75%
Dryas	•	•				spr-aut	15-30	80%
Duranta	•	•				spr-aut	20-50	80%
Dysoxylum	•	•				spr-aut	30-80	40-75%
Eccremocarpus	•	•				spr-aut	15-40	90%
Echinacea					win	spr-aut	10-20	100%
Echinops	spr						20-50	85%
Echium	•	•				spr-aut	25-40	50-75%
Edgeworthia	•	•				spr-aut	25-60	40-80%
Elaeagnus	sum	sum	win			sum-aut	25-75	75%
Elaeocarpus	•	•				sum-aut	30-50	75%
Embothrium		•				sum/aut	30-60	30-60%
Enkianthus	•	•				spr-aut	30-70	30-60%
Entelea	•	•				spr-aut	15-30	90%
Epacris	•	•				summer	30-50	50-75%
Episcia	•	•				spr-aut	15-25	85%
Erica	•	•				spr-aut	25-50	85%
Erigeron	•	•				spr-aut	10-20	100%
Eriobotrya	•	•				sum-aut	30-70	40-75%
Eriostemon				•	any	sum-aut	30-70	60-80%
Erodium	•	•				spr-aut	15-30	80-90%
Eryngium					•	win/spr	30-60	75%
Erysimum	•	•				spr-aut	15-40	80%
Erythrina	•	•				spr-aut	30-70	40-80%
Escallonia	sum	sum	win			spr-aut	25-70	80%
Eucryphia	•	•				spr-aut	30-70	50-75%
Eugenia	•	•				spr-aut	30-70	40-75%
Euonymus	•	•				spr-aut	25-60	75%
Eupatorium	•	•				spring	15-30	90%
Euphorbia	•	•				spr-aut	20-50	75%
Euryops	•	•				any	20-50	30-60%
Eutaxia	•	•				spr-aut	30-70	40-75%
Exochorda	•	•				spr-aut	30-70	90%
Fabiana	•	•				spr-aut	20-40	80%
Fatshedera	•	•				any	15-35	40-75%
Fatsia	•	•				any	20-40	100%
Feijoa	•	•				spr-aut	25-50	70-90%
Felicia	•	•				any	30-70	60-90%
Ficus	sum	sum			win	spr-aut	10-25	50-75%
Forsythia	sum	sum			win	spr-aut	20-60	40-75%
Fothergilla	•	•				spr-aut	25-80	95%
Francoa	•	•				spring	30-90	100%
Fuchsia	•	•				spr-aut	20-40	90%
Gaillardia	•	•				spr-aut	15-30	60-80%
Gamolepis	•	•				spr-aut	10-20	30-65%
Gardenia	•	•				spr-aut	15-30	60-80%
Garrya	•	•				spr-aut	25-50	40-70%
Gaultheria	•	•				spr-aut	25-70	60-80%

Type of Cutting

Stem Cuttings Plant	Soft	Semi	Hard	Leaf	Root	Season	Time (days)	Strike
Gaura	•	•				spr-aut	15-30	60-80%
Gazania	•	•				any	10-25	100%
Gelsemium	•	•				spr-aut	25-60	50-75%
Geniostoma	•	•				spr-aut	20-60	60-80%
Genista	•	•				spr-aut	25-50	60-90%
Gentiana	•	•				spr-aut	20-50	30-75%
Geranium	•	•				spr-aut	10-30	90%
Geum	•	•				spr/sum	15-30	90%
Glechoma	•	•				any	10-25	100%
Goodia	•	•				spr-aut	25-60	30-75%
Gordonia	•	•				spr-aut	30-80	25-75%
Grevillea	•	•				spr-aut	25-60	50-80%
Grewia	•	•				spr-aut	25-60	50-75%
Greyia	•	•				spr-aut	25-60	40-75%
Griselinia	•	•				any	25-60	60-90%
Gypsophila	•	•				spr-aut	20-50	50-80%
Halimiocistus	•	•				spr-aut	30-60	75%
Halimium syn Helianthemum	•	•				spr-aut	20-50	85%
Hamamelis		•				sum/aut	25-70	20-60%
Hardenbergia	•	•				spr-aut	20-50	60-90%
Hebe	•	•	win			spr-aut	25-50	85%
Hedera	sum	sum				any	20-50	95%
Hedycarya	•	•				spr-aut	30-80	95%
Helenium	•	•				spr-aut	15-30	95%
Helianthemum	•	•				spr-aut	15-40	60-90%
Helianthus	•	•				sum/aut	15-30	90%
Helichrysum	•	•				spr-aut	15-30	60-90%
-perennial forms	•	•				spr-sum	10-25	100%
Heliopsis	•	•				spr-aut	15-30	75-90%
Heliotropium	•	•				spr-aut	20-60	60-80%
Heterocentron syn Heeria	•	•				spr-aut	25-60	60-90%
Hibbertia	•	aut				spr-aut	15-30	75%
Hibiscus -hardy	spr	aut				spr-aut	25-60	75%
Hibiscus -tropical	•	•				spr-aut	15-30	50-75%
Hoheria	•	•				spr-aut	20-60	90%
Holmskioldia	•	•				sum/aut	25-60	30-75%
Hovea	•	•				spr-aut	25-80	50-75%
Hovenia	•	•				sum/aut	20-60	40-75%
Hoya	•	•				spr-aut	30-70	75%
Humulus (hop)	•	•				spr-aut	20-60	80%
Hymenosporum	sum	sum	win			spr-aut	30-70	40-75%
Hypericum	•	•				any	20-50	80%
Hypocalymma	•	•				spr-aut	35-80	20-75%

Table 4: Propagation by cuttings

Stem Cuttings (part 1)

Plant	Soft	Semi	Hard	Leaf	Root	Season	Time (days)	Strike
Hyssopus (hyssop)	•					spr-aut	10-25	90%
Iberis (candytuft)								
-perennial	•	•				spr-aut	20-45	70-90%
Ilex	•	•		•		spr-aut	30-80	50-80%
Impatiens	•					spr-aut	10-20	100%
Indigofera	•	•				win/spr	40-75	50-75%
Iochroma	•	•				spr-aut	15-40	70-90%
Ipomoea	•	•				spr-aut	10-30	70-90%
Isoplexis	•	•				spr-aut	25-60	70-90%
Ixerba		•				sum/aut	25-60	75%
Jacaranda	•					spr-aut	20-50	60-75%
Jasminum	•	•				spr-aut	20-50	60-80%
Jovellana		•	•			spr-aut	20-40	85%
Juglans (walnut)						winter	60-100	75%
Juniperus (juniper)								
Justicia syn Beloperone, Jacobinia	•	•				any	30-80	40-75%
Kadsura	•	•				spr-aut	15-35	75-90%
Kalanchoe	•	•				spr-aut	20-50	80%
Kalmia	•		•			spr-aut	15-35	90%
-not latifolia						spr-aut	30-70	30-75%
Kalmiopsis						spr-aut	30-80	50-75%
Kennedia						spr-aut	25-60	60-80%
Kerria	sum	sum	win				25-75	75%
Koelreuteria				•		winter	60-90	75%
Kolkwitzia		sum	win				35-90	80%
Kunzea	sum	sum				spr-aut	25-60	80%
Laburnum			win			winter	60-100	90%
Lagerstroemia	sum	sum					30-100	75%
Lagunaria		•				spr-aut	30-70	50-75%
Lambertia		•				sum/aut	30-70	20-60%
Lamium	•					any	10-30	90%
Lampranthus	•	•				spr-aut	15-40	90%
Lantana	•	•				spr-aut	15-40	90%
Laurelia		•				spr-aut	30-80	60-80%
Laurus		•				sum/aut	30-70	50-75%
Lavandula	•	•				spr-aut	25-70	50-90%
Lavatera	•	•					15-30	90%
-perennial forms	•					spr-aut	15-40	90%
Leonotis	•	•				spr-aut	25-60	80%
Leptospermum		•			win	spr-aut	25-70	20-75%
Leschenaultia		•				sum/aut	30-90	30-75%
Leucadendron		•				spr-aut	25-60	30-75%
Leucopogon		•				spr-aut	25-90	30-75%

Stem Cuttings (part 2)

Plant	Soft	Semi	Hard	Leaf	Root	Season	Time (days)	Strike
Leucospermum		•				sum/aut	30-90	30-75%
Leucothe	•	•				spr-aut	25-70	75%
Levisticum (lovage)								
Liatris	•	•				spr-aut	10-30	90%
Libocedrus		•				spr-aut	15-30	75%
Ligustrum (privet)	•	•				spr-aut	30-100	40-75%
Linum	sum	sum			win			
Liquidambar		sum	win			winter	20-80	90%
Lithodora syn								
Lithospermum				•			15-50	60-80%
Lobelia						winter	60-100	75%
Lomatia								
perennial forms		•				spr-aut	15-40	50-80%
Lonicera	sum	sum	win					
Lophomyrtus		•				spr-aut	10-30	90%
Loropetalum		•				sum/aut	30-90	25-60%
Lotus	•	•				spr-aut	25-80	80%
Luculia		•				spr-aut	30-75	75%
Lupinus (lupin)	•					sum/aut	30-70	40-75%
Lysimachia	•	•				spr-aut	15-40	60-90%
Lythrum	•	•				autumn	30-80	20-60%
Macadamia		•				spring	15-30	90%
Macfadyena syn Doxantha	•	•				any	10-30	100%
Magnolia		•				spr-aut	15-40	70-90%
Mahonia		•				sum/aut	30-90	20-60%
Malus-own roots				•		winter	60-100	75-90%
Malva	•	•				spr-aut	30-90	50-75%
Malvaviscus		•				spr-aut	15-42	40-80%
Mandevilla		•				spr-aut	20-70	50-80%
Manettia	•	•				spr-aut	14-35	70-90%
Marianthus		•				spr-aut	15-42	90%
Marrubium (horehound)		•				spr-aut	20-70	40-80%
Maytenus		•				spr-aut	15-40	100%
Mazus	•	•				spr-aut	20-50	60-80%
Melaleuca		•				spr-aut	15-30	90%
Melia		•		•		spr-aut	30-70	50-75%
Melicope		•				spr-aut	15-40	70-90%
Melicytus		•				spr-aut	30-70	40-80%
Melissa (balm)	•					spr-aut	25-80	60-80%
Mentha (mint)	•	•				spr-aut	25-70	60-80%
Meryta		•			•	any	10-30	90%
Mesembryanthemum (ice plant)		•				sum/aut	10-25	100%
Metasequoia	•	•				spr-aut	15-40	75-90%
Metrosideros		•				sum/aut	25-60	50-80%
		•				spr-aut	30-80	40-80%

81

Type of Cutting — Stem Cuttings (Soft / Semi / Hard), Leaf, Root

Plant	Soft	Semi	Hard	Leaf	Root	Season	Time (days)	Strike
Michaelia	•	•				spr-aut	25-70	50-75%
Micromyrtus	•	•				spr-aut	25-60	50-75%
Mimetes		•	•			sum/aut	30-80	30-60%
Mimulus	•	•				spr-aut	15-30	90%
Mitraria	•					spr-aut	15-40	90%
Monarda (bergamot)	•					spr-aut	10-30	100%
Monstera		•				sum/aut	20-60	80%
Morus (mulberry)			•			winter	75-100	75%
Moschosma	•					spr-aut	15-40	90%
Muehlenbeckia	•					spr-aut	25-60	75%
Murraya		•				spr-aut	25-70	50-75%
Mussaenda	•					spr-aut	25-60	50-80%
Mutisia		•				spr-aut	15-50	70-90%
Myoporum		•				spr-aut	15-40	90%
Myosotis								
-perennial forms	•					sum/aut	15-40	90%
Myrsine		•				spr-aut	15-40	90%
Myrtus		•				spr-aut	25-60	60-80%
Nandina		•				sum/aut	25-60	60-85%
Nemesia								
-perennial forms	•					spr-aut	10-40	90%
Nepeta	•					any	10-30	90%
Nerium		•	win			sum/aut	35-80	50-75%
Nierembergia	•					spr-aut	15-40	90%
Notospartium	•					sum/aut	20-60	30-75%
Odontospermum	•					spr-aut	14-35	70-90%
Oenothera (evening primrose)					win			
Olea (olive)		•				spr-aut	15-40	60-80%
Olearia	•	•				sum/aut	40-90	40-75%
Origanum	•					spr-aut	25-70	60-90%
Osmanthus		sum	win			spr-aut	15-30	100%
Osmarea		sum	win			spr-aut	30-100	50-75%
Osteospermum	•							
Oxypetalum syn Tweedia	•					spr-aut	15-40	50-75%
Pachysandra	•					spr-aut	15-40	70-90%
Pachystegia	•					any	20-60	85%
Pandorea		•				spr-aut	25-50	60-85%
Parahebe	•					spr-aut	30-70	60-85%
Parsonsia		•				any	15-30	90%
Parthenocissus		sum	win			spr-aut	30-80	40-75%
Passiflora		•				spr-aut	25-70	90%
Pelargonium	•					spr-aut	15-40	90%
Penstemon	•					spr-aut	15-40	85%
Pentas	•	•				spr-aut	25-50	50-75%
Pernettya		•				spr-aut	25-60	70-90%

Plant	Soft	Semi	Hard	Leaf	Root	Season	Time (days)	Strike
Perovskia	•	•				spr-sum	20-40	60-80%
Persoonia	•	•				sum/aut	30-70	25-60%
Phaenocoma		•	•			spr-aut	15-35	90%
Phebalium		•				sum/aut	25-60	70-90%
Philadelphus		•				sum/aut	30-90	60-80%
Philodendron -climbing		sum						
Phlomis	•	•				spr-aut	15-40	70-90%
Phlox -rockery and trailers	•					spr-aut	15-30	90%
Phlox paniculata					•	spr-aut	15-40	70-90%
Photinia		•				sum/aut	40-80	80%
Phygelius	•					aut-win	30-70	60-80%
Phylica		•				sum/aut	15-30	90%
Physostegia	•					any	30-70	80-95%
Pieris		•				sum/aut	15-30	60-80%
Pimelea		•		•		spr-aut	30-60	60-80%
Pisonia		•				spr-aut	30-60	70-90%
Pittosporum		•				sum/aut	25-60	50-80%
Plagianthus		•				spr-aut	30-100	40-70%
Platanus			win			winter	70-100	80%
Platycodon	•					spring	15-30	85%
Plectranthus	•	•				spr-aut	15-40	90%
Plumbago		•				spr-aut	25-60	60-90%
Plumeria (frangipani)		•				winter	50-100	50-75%
Podalyria		•				sum/aut	40-80	20-60%
Podocarpus		•				any	40-100	30-75%
Podranea		•				spr-aut	15-40	90%
Polemonium (Jacob's ladder)	•					spr-aut	10-30	100%
Polygala	any		win			spr-aut	30-70	50-75%
Polygonum	•					sum/aut	15-40	90%
Pomaderris		•				spr-aut	25-60	60-80%
Populus (poplar)			•			winter	60-100	90%
Portulacaria		•				spr-aut	15-30	90%
Potentilla	•					spr-aut	25-70	90%
Pratia	•					spr-aut	15-30	90%
Prostanthera		•				spr-aut	30-70	60-80%
Protea		•				sum/aut	40-90	20-75%
Prunus laurocerasus		•				spr-aut	28-60	70-90%
Pseudopanax		•				sum/aut	25-70	60-80%
Pseudowintera		•				spr-aut	30-80	30-65%
Psoralea		•				spr-aut	20-60	70-90%
Pterocarya					•	winter	60-100	75%
Pulmonaria	•					spr-aut	15-30	90%
Pulsatilla				•		winter	40-80	85%

Table 4: Propagation by cuttings

Plant	Stem Cuttings — Type of Cutting					Season	Time (days)	Strike
	Soft	Semi	Hard	Leaf	Root			
Punica	•	sum	win			any	30-80	75%
Pyracantha	•	•				spr-aut	25-60	70-90%
Pyrostegia		•				spr-aut	20-60	60-80%
Quintinia		•				sum/aut	30-80	50-75%
Raphiolepsis	•	•				spr-aut	30-70	60-80%
Regelia	•	•				spr-aut	20-50	60-80%
Reinwardtia	•	•				spr-aut	20-60	70-90%
Rhabdothamnus		•				sum/aut	20-60	80%
Rhamnus		•				spr-aut	30-70	50-75%
Rhododendron -small leaf		•				spr-aut	30-70	60-90%
Rhododendron -dec. azalea	•					spring	40-70	30-80%
Rhododendron -ev. azalea		•				spr-aut	30-70	60-95%
Rhododendron -large leaf		•				sum/aut	40-100	30-90%
Rhododendron -vireya	•					spr-aut	40-80	50-90%
Rhus		sum		•	•	winter	40-80	90%
Ribes		•	win	•	•	winter	30-100	80%
Robinia-species				•	•	win-spr	60-100	30-60%
Romneya	•	•				spr-aut	40-100	90%
Rondeletia	•	•				spr-aut	15-40	
Rosa-miniature	•	•				spr-aut	20-50	70-90%
Rosa -non budded, own roots	sum	sum	win			spr-aut	30-80	40-90%
Rosmarinus	sum	win				spr-aut	25-60	60-80%
Rubus	sum	win			win	spr-aut	25-80	90%
Rudbeckia	•	•				spr-aut	10-30	100%
Ruscus	sum	sum				spr-aut	40-80	30-60%
Russelia	•	•				spr-aut	20-60	70-90%
Ruta (rue)	•	•				spr/sum	10-30	95%
Salix			win			winter	40-80	100%
Salvia -perennial forms	•	•				spr-aut	15-30	90%
Sanseveria	•	•				spr-aut	21-42	90%
Santolina	•	•				spr-aut	20-40	90%
Sapium	•					winter	60-100	75%
Saponaria (soapwort)	•	•				spr-aut	15-30	90%
Sarcococca	•	•				spr-aut	30-70	40-75%
Satureja (savoury)	•	•				spr-aut	10-30	90%
Scabiosa	•	•				spr-aut	20-50	40-80%
Schefflera		•				spr-aut	20-50	90%
Schinus		•				spr-aut	30-70	40-75%
Sedum	•					any	10-30	100%

Plant	Stem Cuttings — Type of Cutting					Season	Time (days)	Strike
	Soft	Semi	Hard	Leaf	Root			
Selago	•	•				spr-aut	15-40	90%
Senecio	•	•				any	20-60	70-90%
Sequoiadendron -dwarf forms		•				any	30-90	50-75%
Serissa	•	•				spr-aut	15-40	90%
Serruria		•				autumn	30-80	20-60%
Shortia	•	•				spr-aut	30-80	60-80%
Silene	•	•				spr-aut	15-40	70-90%
Skimmia	sum	sum				spr-aut	30-90	75%
Solandra	sum	•				spr-aut	20-50	85%
Solanum	•	•				spr-aut	15-50	90%
Soleirolia (baby's tears)	•	•				any	5-15	100%
Sollya	•	•				spr-aut	25-60	60-80%
Sophora		•	win			win/spr	30-100	30-75%
Sorbus		•	win			winter	50-100	25-60%
Sparmannia	•	•				any	25-60	85%
Spartium	sum	sum	win			spr-aut	30-90	60-80%
Spathodea		•				sum-aut	25-70	80%
Spiraea	sum	sum	win			spr-aut	25-90	80%
Stachyurus	sum	sum	win			winter	25-50	75%
Staphylea		•				spr-aut	30-90	80%
Stauntonia		•				spring	40-100	20-60%
Stephanandra		•	•			spr-aut	50-100	80%
Stephanotis		•				spr-aut	35-90	25-60%
Stewartia		•	•			spr-aut	30-70	25-60%
Stokesia		•				spr-aut	15-30	90%
Stranvaesia		•				spr-aut	25-60	60-80%
Streptocarpus		•				spr-aut	20-50	100%
Streptosolen		•				spr-aut	20-60	70-90%
Swainsona		•				spr-aut	15-50	80%
Symphoriocarpos		•	•			winter	50-100	80%
Symphytum (comfrey)					win	spr-aut	15-30	90%
Syringa	•	sum			win	spr-aut	30-90	50-80%
Syzygium		•				spr-aut	25-60	60-80%
Tamarix	sum	sum	win			winter	60-100	80%
Tanacetum (tansy)	•	•				spr-aut	10-30	100%
Taxus (yew)	sum	sum	win			spr-aut	30-150	40-75%
Tecoma	•	•				spr-aut	15-40	90%
Tecomanthe	•	•				spr-aut	25-60	75%
Tecomaria	•	•				sum/aut	20-60	70-90%
Telopea	•	•				spr-aut	30-90	20-60%
Templetonia	•	•				spr-aut	25-60	75%
Ternstroemia	•	•				sum/aut	30-70	75%
Tetrapanax	•	•				spr-aut	21-42	90%
Tetrapathea	•	•				sum/aut	25-60	80%
Teucrium	•	•				any	30-70	60-80%

Type of Cutting

Stem Cuttings								
Plant	Soft	Semi	Hard	Leaf	Root	Season	Time (days)	Strike
Thryptomene		•				sum/aut	30-70	60-80%
Thuja		•				aut-win	60-150	75%
Thujopsis		•				aut-win	60-100	80%
Thunbergia	•	•				spr-aut	15-40	70-90%
Thymus (thyme)		•				any	15-30	100%
Tibouchina	•					spr-aut	25-70	40-75%
Tilia (lime)		•				autumn	35-80	40-60%
Tolmiea				•		any	10-20	100%
Toronia		•				sum/aut	40-100	20-60%
Trachelospermum		•				spr-aut	30-70	40-75%
Tradescantia	•					any	10-30	100%
Tristania		•				sum/aut	25-70	40-80%
Tsuga		•				any	40-100	75%
Ugni syn	•					spr-aut	25-60	70-90%
Myrtus ugni			•			winter	60-100	75%
Ulmus (elm)	•					spr-aut	30-80	80%
Vaccinum	•					spr-aut	15-30	90%
Valeriana		•				spr		
Verbascum					win		20-60	90%

Type of Cutting

Stem Cuttings								
Plant	Soft	Semi	Hard	Leaf	Root	Season	Time (days)	Strike
Verbena								
-perennial forms	•					any	10-30	100%
Veronica	•					spr-aut	10-30	100%
Vestia		•				spr-aut	15-40	90%
Viburnum		sum	win				30-80	60-80%
Vigna syn								
Phaseolus	•	•				spr-aut	15-40	90%
Vinca	•					any	10-30	100%
Viola	•	•				any	10-30	100%
Virgilia		•				sum/aut	30-70	40-60%
Vitex		•				spr-aut	30-80	40-75%
Vitis (grape)	sum		win				20-80	90%
Weigela	sum	sum	win				25-80	85%
Weinmannia		•				spr-aut	30-80	50-75%
Westringia	•	•				spr-aut	30-70	60-80%
Wisteria		aut	win				40-100	40-75%
Zauschneria	•	•				spr-aut	15-40	90%
Zenobia		•				spr-aut	30-80	50-75%

Chapter 11

LESS COMMON PROPAGATION TECHNIQUES

Rhododendron 'Anna-Rose Whitney'

Layering

LAYERING is a development of a natural process that is most commonly seen in ground covers, which often form roots at the points where their stems touch the ground. You can simulate this natural layering by keeping a branch in contact with the soil. With time, roots will form and the struck layer may be separated from its parent plant. Most shrubs and trees can be layered, provided they have branches or stems close to the ground.

Greater success will be achieved if the stem is wounded before you bury it. Make a shallow cut into the branch, being careful not to seriously weaken it. Dab a little rooting hormone on the wound then peg the stem to the ground with a hoop of wire and mound soil over the wounded area. If it is possible to scratch out a small trench for the stem, so much the better, as mounded soil may blow away.

Layering is a near foolproof propagation method, but it has several drawbacks. Layers are generally not quick to strike (the time varies depending on the season and the plant involved, but between 9–12 months is typical); unless you can stake the growing tip of the layer as it develops you may find that it has lopsided growth; and layering is

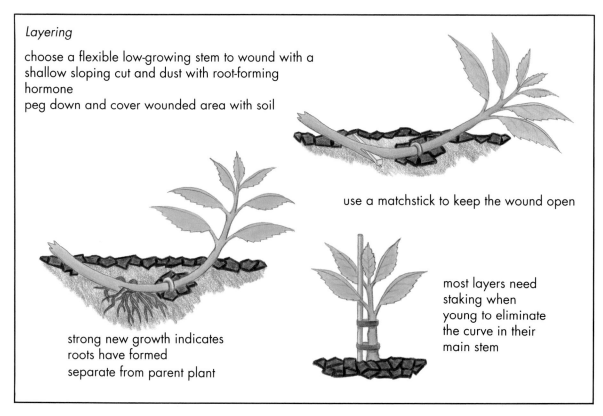

Layering

choose a flexible low-growing stem to wound with a shallow sloping cut and dust with root-forming hormone
peg down and cover wounded area with soil

use a matchstick to keep the wound open

strong new growth indicates roots have formed
separate from parent plant

most layers need staking when young to eliminate the curve in their main stem

also limited to fairly small-scale production.

The advantages of layering are that it does not involve the removal of material from the parent plant until it has struck and that it can be performed *in situ*, without a greenhouse or other environmental controls. If you want to propagate a rare plant that provides limited cutting material or a plant that is hard to strike from cuttings and difficult to graft, layering may be the answer.

There are many variations on the layering process that are really just adaptations to suit different plant structures.

Tip layering

Raspberries, blackberries and other plants with long, whippy, cane-like stems will strike reasonably quickly if layered at the extreme tip of the stem. The process is the same as regular layering except that just the tip of the branch is pegged to the soil and buried at the point of contact. Tip layers strike in about 6–8 weeks if taken as soon as the

canes are long enough to be bent to ground level.

Stooling

Stooling or mound layering works best with deciduous shrubs and is particularly useful when the stems are too inflexible to be bent down to ground level.

The parent plant should be cut back very hard in late winter (deciduous plants may be cut back almost to ground level). This hard pruning will cause strong-growing basal shoots to form in spring. As the shoots grow, soil is mounded up around them, leaving just the growth tips exposed. This procedure is repeated until late summer, after which the stems are allowed to grow as normal until winter. When the plant is once again dormant, the soil can be removed and you should find that the stems have developed roots where they have been buried. The newly formed layers can be removed and potted on or planted out. The parent plant

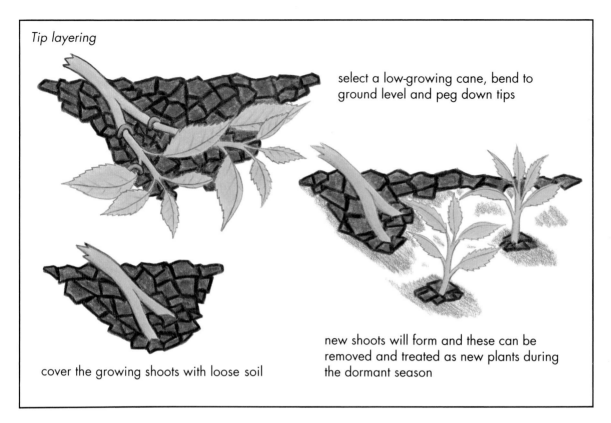

Tip layering

select a low-growing cane, bend to ground level and peg down tips

cover the growing shoots with loose soil

new shoots will form and these can be removed and treated as new plants during the dormant season

can be cropped in this way for some years.

When stooling evergreens, the pruning should not be quite so severe. Many shrubs, especially callunas, European ericas and dwarf rhododendrons, have masses of thin wiry shoots at ground level. If the plants are cut back hard in late winter or early spring and soil is mounded up around the plant as the new shoots grow, most of the new shoots will layer and can be removed the following autumn.

Trench layering

Trench layering, or French layering, is a variation of stooling. The plants are not hard pruned, but very twiggy bushes should be lightly cut back and thinned in winter. About 8–12 60-cm branches is a manageable size and number. In late winter prepare shallow trenches around the parent plant, bend down the stems and peg them to the ground then mound them over with soil.

As the stems grow, new shoots forming

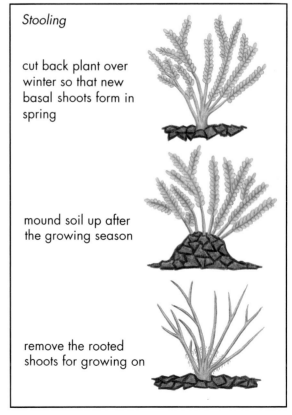

Stooling

cut back plant over winter so that new basal shoots form in spring

mound soil up after the growing season

remove the rooted shoots for growing on

Trench layering

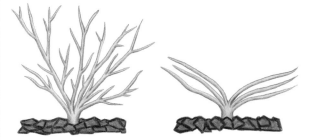

in spring, new shoots will appear

early in growing season cut back plant to encourage new shoots
at the end of the growing season cut back to 6–8 shoots that show horizontal growth

mound soil around shoots so that only the tips show

bend to ground level and peg down tip growth, covering with fine soil

uncover shoots in the following season and separate from parent plant

along the branches will push up through the soil. These too are mounded up with soil, leaving just the tips exposed. The following dormant season the whole length of the original branch and its side shoots can be unearthed and divided up.

Dropping

Dropping is a further variation on stooling, but instead of mounding the soil up around the plant, the plant is buried. It works best with shrubs that develop masses of thin shoots, such as ericas and callunas. The plant is pruned, either in winter or early spring, to encourage a large number of shoots. An area of soil is prepared nearby by adding leaf mould, peat or the like so that it is loose and airy. The plant is then lifted and buried, angled on its side with just the leaf tips exposed, in the specially prepared soil. By the following winter the plant will have produced numerous struck layers, which may be removed and grown on.

Aerial layering

When you can't bend a branch down to the soil and the plant is too large for stooling, the only way to layer it is to take the soil to the branch. This is effectively how aerial layering works, and it is a useful technique for plants with large leaves and heavy stems, such as rubber plants, that are rather unwieldy to handle as cuttings.

Aerial layering does not require any specialised equipment but you will need a sharp knife, some wet sphagnum moss, a patch of black polythene sheeting about 30 cm square, a little root-forming hormone (not essential), and some wire ties.

It is best to work with the current year's growth. Choose a reasonably short stem, say about 45 cm long, or work near the tip of a branch. The wood should be firm but not completely hardened. Harder wood will strike but it takes considerably longer and the root structure is often poor.

Trim the foliage just above and below the

area that you intend to work on so that you have easy access to the stem. Slit the bark about half way round the stem. Make an identical cut about 5 mm above or below the first and remove the strip of bark in between. This will stop the wound from entirely healing over. Do not completely ring-bark the stem. If you are working with a fairly soft stem, remove a sliver of bark by making a shallow cut along the stem.

Having made the cut, lightly dust the wound with root-forming hormone powder and wrap the stem with wet sphagnum moss. Hold the sphagnum in position while you wrap the polythene around it and then tie the polythene at each end, thereby securing the ball of sphagnum. It is usually easiest to have someone help you with this part.

The moss will keep the wound from drying out and healing over, just like the soil in a conventional layer, with the end result that roots form at the wound. The plastic in turn prevents the moss from drying. Black polythene is best because it blocks out light and raises the temperature considerably.

Aerial layering is usually done in mid to late summer, with the plants striking the following spring. However, if you start earlier, say mid spring, it is possible to have a strike in as little as 6–8 weeks. Check the layer by carefully unwrapping the polythene. When it is apparent that a good root system has formed, the layer may be removed and potted up for growing on.

Grafting

Grafting is probably the most advanced propagation method that can be done without very sophisticated equipment. It is mainly used for propagating selected forms of ornamental trees and fruit trees. It is a fairly common practice in nurseries, but few home gardeners graft plants.

A grafted plant is composed of a stock, which is a plant on which to graft, and a scion, which is the wood that is to be graft-

Grafted rhododendrons sitting on gravel in a container with a few centimetres of water. The container is usually covered with glass to create a very humid atmosphere.

ed. All grafting methods require the cutting and trimming of stock and scion into shapes that will enable them to be neatly joined, so a sharp knife and steady hands are essential. Once joined, the scion is taped in place until it has taken. You will find that successful grafting requires regular practice and that it is a time-consuming process, at least at first. It will soon become clear to you why grafted plants are usually the most expensive you can buy.

Grafting aims to match the cambium layers of two different plants so that they eventually fuse together to become one. The plants must be reasonably closely related, and even then they may not be completely compatible. Also, there may be problems with varying growth rates between the stock and scion, and with stocks that sucker badly.

A successful saddle graft on a rhododendron.
Note the callus.

The best time for grafting is in spring, as the sap is rising, but if the scion material comes into leaf too early it will collapse before the graft is united. Consequently, scion wood is usually collected in late winter, packed so that it remains moist, and kept in cool storage until needed. Although this is fine with deciduous plants it isn't practical with evergreens. One way around this problem is to pot up the stock and keep it indoors over winter so that it comes into spring growth earlier than normal. The scion donor is kept outdoors so that it remains dormant, and the graft is performed once the stock is growing strongly.

Many grafting methods have been developed, but relatively few are regularly used. The whip and tongue graft is used when the stock's stem is less than 20 mm in diameter. It is quite a reliable method but ruins the stock as a useful plant in its own right, so if anything goes wrong all is lost. It also requires that the stock and scion be the roughly the same diameter to ensure a close cambium match.

Side grafts are not so restricting because the stock can be used again and the stock and scion can be different sizes, but there is the drawback that the grafted branch will always grow at a slight angle.

Approach grafting, which most closely simulates the natural grafting that may occasionally be seen when two branches remain in contact for a prolonged period, is usually a last resort. That's not because it is any less successful than any other means, but because it requires two or more plants to be kept in precise positions until the graft union is well established. Any disturbance before the graft union is established and the graft is ruined.

The best method for home gardeners is usually the saddle graft, or its inverted version, the apical wedge. Cutting the stock and scion with these methods requires no great carpentry skill and they work well with a wide range of plants. Generally, saddle grafts should be used for evergreens, such as rhododendrons, while the apical wedge works better with deciduous material.

Freshly grafted plants perform best if they can be kept in a warm humid environment. This is especially true of evergreens, which may wilt and suffer fatal moisture stress before the graft can take. There are two commonly used ways of keeping the graft area moist, although they are only effective with potted plants. The first is to plunge the entire plant in moist sawdust until the scion is growing well. The second method is to line the bottom of a trough or tub with a few centimetres of gravel and then pour in water until it is just level with the surface of the gravel. The pots can sit on the gravel and the water will ensure the atmosphere remains humid. It the trough can be covered with a few panes of glass, so much the better.

Once the scion is growing well and the graft has callused over, the tape can be removed. Don't leave this too long or you may find the tape starts to cut into the stem.

Budding

Budding is most commonly used with roses, but it is equally successful with many of the wider rose family members, such as apples, quinces, pears and the *Prunus* species. Most plants that can be grafted may also be budded, but budding has one major disadvantage for ornamental stock: a bud is attached to the side of a stem so there is always a tendency for it to grow outwards before it heads up.

This is a disadvantage because it puts more weight on the bud union. As this will be a weak point for quite some time, there is an increased risk of damage, particularly in high winds or if excessive growth is allowed to develop on a newly budded shoot. On the other hand, it is not necessary to destroy the stock plant in order to insert a bud, so if a bud fails to take, nothing is lost.

Shield budding

Shield budding, also known as 'T' budding, is the most common technique, and is the one usually used for roses. The best time for budding is from late spring to late summer when the sap is flowing and the bark is soft and easy to work with, but it can also be done in late winter and early spring.

To produce top-quality plants, you need the best rootstocks. Specialist nurseries grow large quantities of stock plants and can usually be persuaded to part with some, provided you take a decent number. One-year-old stocks are best; plant them out in winter to give them time to get established. However, if you just want to have a go at budding to see how it's done, you can use your existing plants as rootstocks.

Start by preparing the rootstock. If you are using a purpose-grown stock you should work close to the ground, but if you intend to bud onto existing plants you can bud higher up. However, the closer to the roots the bud is, the less chance there is of stock regrowth.

The bark of the stock should be soft as it must lift easily when cut, but it shouldn't be so soft and green that it starts to wither and

dry immediately it is cut. That's why young stocks are best: they still have soft bark near ground level. Trim away any foliage where you intend to bud, then, using a very sharp knife, make a T-shaped cut. The exact size of the cut varies with the size of the bud, typically the stem of the 'T' will be a cut about 25 mm long, while the cross cut will be about 15–20 mm. Make sure that you don't ring bark the stem or cut deeply into the wood.

Next, prepare the bud. Look for a mature leaf with a firm bud at the leaf axil. The bud should be plump and fleshy without showing any signs of starting into growth. Remove the leaf, leaving about 10–15 mm of the leaf-stalk. Next remove the bud by making a shallow cut under it. Start about 1 cm below the bud, pass under it and continue the cut for about 20 mm above the bud. This will remove a sliver of bark and stem with the bud and leaf-stalk attached. Clean off any parts of woody stem attached to the bark and your bud will be ready to insert.

Gently lift the bark away from the T-shaped cut on the stock, then slide the bud-wood under the bark until the bud is just below the level of the cross cut. Close the flaps of bark back over the bud and trim the surplus bud-wood bark level with the cross cut. Tie the bud using grafting tape, ordinary Sellotape or plumbers' thread tape, leaving the bud exposed. Rose growers often use specialist budding patches that cover the entire bud.

You will know if the bud has taken within a few weeks. If the stub of petiole drops, leaving the bud looking healthy, then the chances are that you have been successful. The wound will heal quickly but the bud will not start to grow until the following spring. The tape can be removed and the stem above the bud cut back in late winter or early spring, before growth starts.

In areas of high rainfall, or if the stock plant is inclined to bleed heavily when cut,

Length of stem and bud before preparation. (For clarity, this is a little larger than would normally be used.)

Rootstock prepared for budding. (The slivers are holding the cut open to make it easier to photograph, they are not normally used.)

Bud ready for insertion. The leaf will drop off naturally and in the meantime can be used as a handle.

Bud inserted in the cut and ready for binding.

Bud firmly bound, but with room for some air and light to penetrate.

the incision is sometimes made the other way up, with the top of the 'T' at the base. This will allow rainwater and excess sap to run off, but there is the risk that the bud may fall out or move within the cut.

Chip budding

Chip budding is generally only used when shield budding is not practical, such as with plants that have brittle bark or bark that is difficult to separate from the stem. You may chip bud in summer, autumn or winter and you should find that chip buds start to grow sooner than shield buds. However, the failure rate is higher and the union tends to be weaker.

The first step in chip budding is to prepare the stock to receive the bud. First, using a very sharp knife, make a cut across and slightly downward into your stock stem. This cut should be at a shallow angle toward the centre but should only be about 5 mm deep. Next, starting 30 mm above your first cut, make a shallow cut along the stem. This should intersect the first cut so that you can remove a wedge or chip of bark and stem. The second cut should be angled so a notch is left when the chip is removed.

Now, working with the scion, cut an identically shaped chip, with a bud at its centre. Insert this chip into the cut on the stock. Make sure that it is locked into the notch at the base of the cut and that the cambium layers match, then tape the chip into position.

Root cuttings

Some plants may be grown from small lengths of root. This is now an uncommon method of propagation because it takes a considerable time and may require preparation well in advance of taking the cuttings.

Start by lifting the plant when it is dormant to establish what sort of root structure it has. If there are a number of large fleshy roots that are branching well, then you may take cuttings immediately. If not, you will need to prepare the plant by trimming the roots back to within about 100 mm of the crown and then replanting. Strong new roots should develop during the next growing season.

When the roots are ready, lift the plant and remove any excess top growth so that it is easy to handle. Wash the root system, and remove the young fleshy roots near the crown. The original stock may now be replanted. Trim any fine side shoots away from the roots you have removed so that you have a clean length to work with. The interior of the roots should be white, not brown or discoloured. If the roots appear at all diseased, do not use them.

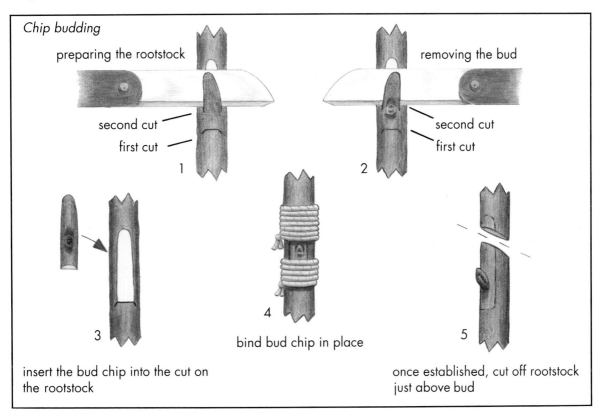

Chip budding

preparing the rootstock

removing the bud

second cut

first cut

1

second cut

first cut

2

3

insert the bud chip into the cut on the rootstock

4

bind bud chip in place

5

once established, cut off rootstock just above bud

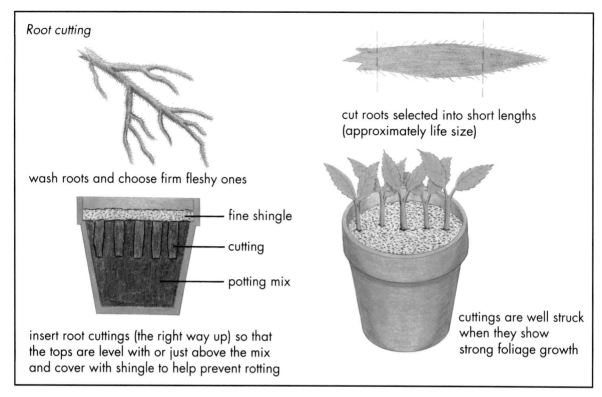

Root cutting

wash roots and choose firm fleshy ones

cut roots selected into short lengths
(approximately life size)

— fine shingle

— cutting

— potting mix

insert root cuttings (the right way up) so that
the tops are level with or just above the mix
and cover with shingle to help prevent rotting

cuttings are well struck
when they show
strong foliage growth

Now trim the roots into pieces of the right length, which is about 100–150 mm if the cuttings are to be kept outdoors, and 50 mm or so for those in heated propagating units.

Use a tray or box for your root cuttings. Fill the box to about halfway with a good sterile potting or cutting mix. Insert the cuttings without pushing them all of the way in, as they can be covered when you are finished. Root cuttings are inserted vertically. As there is no easy way of telling which is the top (and this is important when planting your cuttings), make sure that you always cut from the top towards the tip, and insert the roots as you go. Once the cuttings are all in, fill the box with soil so that the tops of the roots are just level with the soil surface. Top up the box with another 10 mm of river sand, perlite or fine pebbles. The topping keeps the roots in the dark and prevents them becoming too wet. Do not use root-forming hormones on root cuttings.

Some plants, such as *Acanthus*, are less fussy and will sprout from any piece of root

that is kept in moist potting mix.

You will know the cuttings have struck when leaves start to appear. Allow some good shoots to develop before carefully lifting the cuttings for potting on or planting out.

Tissue culture

Tissue culture is the most sophisticated form of propagation. All methods rely on meristems for their success, but tissue culture is the only one to use them directly. For this reason it is sometimes known as meristem culture or cell culture.

Tissue culture involves dissecting a growth tip and removing the meristem tissue. This must be done under sterile laboratory conditions. The meristem is then induced into growth by the careful use of various phytohormones, which initiate the development of roots and leaves. For now, tissue culture is restricted to the laboratory or to special units for very keen propagators, but in time it may be able to be used more widely.

INDEX